THE MINIATURE FRUIT GARDEN; OR, THE CULTURE OF PYRAMIDAL AND BUSH FRUIT TREES, WITH INSTRUCTIONS FOR ROOT-PRUNING &C

THE MINIATURE FRUIT GARDEN; OR, THE CULTURE OF PYRAMIDAL AND BUSH FRUIT TREES, WITH INSTRUCTIONS FOR ROOT-PRUNING &C

Thomas Rivers

www.General-Books.net

Publication Data:

Title: The Miniature Fruit Garden
Subtitle: Or, the Culture of Pyramidal and Bush Fruit Trees, With Instructions for Root-Pruning c
Author: Thomas Rivers
General Books publication date: 2009
Original publication date: 1873
Original Publisher: Longmans, Green
Subjects: Fruit-culture
Gardening / General
Gardening / Fruit
Gardening / Vegetables
Nature / Trees Forests
Technology Engineering / Agriculture / General
Technology Engineering / Agriculture / Agronomy / Crop Science

CONTENTS

1

SECTION 1

INTRODUCTION.

My attention was drawn to the benefits fruit trees derive from root-pruning and frequent removal about the year 1810. I was then a youth, with a most active fruit-appetite, and if a tree bearing superior fruit could be discovered in my father's orchard-like nursery, I was very constant in my visits to it.

In those days there was in the old nursery, first cropped with trees by my grandfather about the middle of the last century, a ' quarter ' – *i. e.,* a piece of ground devoted to the reception of refuse trees – of such trees as were too small or weak for customers ; so that in taking np trees for orders during the winter they were left, and, in spring, all taken up and transplanted to the ' hospital quarter,' as the labourers called it. The trees in this quarter were often removed, – they were, in nursery parlance, ' driven together ' when they stood too thinly in the ground; or, in other words, taken up, often annually, and planted nearer together, on the same piece of ground. 'This old nursery contained about eight acres, the soil a deep reddish loam, inclining to clay, in which fruit trees flourished and grew vigorously. I soon found that it was but of little use to look among the young free-growing trees for fruit, but among the refuse trees, and to the ' hospital quarter' I was indebted for many a fruit-feast – *such* Ribaton Pippins! *such* Golden Pippins !

When I came to a thinking age, I became anxious to know why those refuse trees never made strong, vigorous shoots, like those growing in their immediate neighbourhood, and yet nearly always bore good crops of fruit. Many years elapsed before I saw ' the reason why,' and long afterwards I was advised by a friend, a F. H. S., to write a crude, short paper on the subject, and send it to be read at a meeting of the Horticultural Society: this paper, is published in their ' Transactions.' I had then practised it several years ; so that I may now claim a little more attention, if the old adage that ' practice makes perfect' be worthy of notice.

This little work is not designed for the gardens and gardeners of the wealthy and great, but for those who take a personal interest in fruit-tree culture, and who look on their garden as a never-failing source of amusement. In some few favoured districts, fruit trees, without any extra care in planting and after- management, will bear good crops, and remain healthy for many years. It is not so in gardens with unfavourable soils : and they are greatly in themajority. It is to those possessing such, and more particularly to the possessors of small gardens, that the directions here given may prove of value. The object constantly had in view is to make fruit trees healthy and fruitful, by keeping their roots near the surface. The root-pruning and biennial or occasional removal, so earnestly recommended, are the proper means to bring about these results, as they place the roots within the influence of the sun and air. The ground over the roots of garden trees as generally cultivated is dug once or twice a-year, so that every surface-fibre is destroyed and the larger roots driven downwards ; they, consequently, imbibe crude, watery sap, which leads to much apparent luxuriance in the trees. This, in the end, is fatal to their well-doing, for the vigorous shoots made annually are seldom or never ripened sufficiently to form blossom-buds. Canker then comes on, and, although the trees do not die, they rarely give fruit and in a few years become victims of bad culture, existing in a sort of living death.

There is, perhaps, no fruit tree that claims or deserves our attention equal to a pear. How delicious is a fine melting pear all the winter months ! and to what a lengthened period in the spring may they be brought to table ! Till lately, Beurre Ranee has been our best spring pear ; but this is a most uncertain variety, rarely keeping till the end of May, and often ripening in January and February.

The Belgian pears, raised many years since by the late Major Esperen, and more recently by Monsieur Gregoire, are likely for the present to be the most valuable for prolonging the season of rich melting pears; and of these Josephine de Malines and Bergamotte d'Esperen are especially deserving of notice: they have the excellent quality of ripening slowly. But improvement will, I have no doubt, yet take place ; for pears are so easily raised from seed, and so soon brought into bearing by grafting or budding them on the quince stock, that new and valuable late pears will soon be as plentiful as new roses.

In the following pages it will be seen that I strongly advocate the culture of pyramidal fruit trees. This is no new idea with me. I have paid many visits to the Continental gardens during the greater portion of my active life in business, and have always admired their pyramidal trees when well managed, and I have for many years cultivated them for my amusement; but, owing to a seeming prejudice against them amongst some English gardeners, I was for some time deterred from recommending them, for I thought that men older than myself must know better ; and when I heard some of our market-gardeners and large fruit-growers in the neighbourhood of London scoff at pears grafted on the quince stock as giving fruit of a very inferior flavour, I concluded, like an Englishman, that theforeigners were very ignorant, and very far behind us in the culture of fruit trees.

It was only by repeated visits to foreign gardens that this prejudice was dispelled. I felt convinced that our neighbours excelled us in the management of fruit trees adapted to the open borders of our gardens. I have therefore endeavoured to make the culture of pyramidal trees easy to the uninitiated ; and, having profited largely by experience in attending to it with my own hands, I trust that my readers will benefit by the result.

A humid, mild climate seems extremely favourable to the well-doing of the pear on the quince stock. Jersey, with its moist warm climate, as is well known, produces the finest pears in Europe : these are, for the most part, from trees on quince stocks. The western coast of Scotland, I have reason to know, is favourable for the culture of pear trees on the quince ; and within these very few years Ireland has proved remarkably so, more particularly in the south, where some of our finest varieties of pears on quince stocks are cultivated with perfect success.

PREFACE

THE SIXTEENTH EDITION.

In giving the seventeenth thousand of iny little book to the public, I trust I may be allowed to express my pleasure and gratitude for its success, – perfectly unprecedented in books devoted to horticulture. The reception given to it by those numerous and increasing horticultural amateurs who seem to love to devote their leisure to the culture of fruit and fruit trees, has been to me a source of much pleasure. For some twenty-five or thirty years have I watched the growth of this taste in England, and more particularly in those who garden with their own hands and heads: it is such men that form the true vanguard of fruit culturists, for they almost invariably improve on any suggestidn given by a writer ; and, if 1 wanted them, I could fill a volume with letters from clever amateurs who have given new ideas, always suggestive if not always practicable. As a prominent but not new feature in this enlarged edition, I may refer to the management, and above all the protection, of low lateral cordonfruit trees. I have also pointed out more forcibly than in former editions the capability of growing choice pears and apples on any low cheap walls, and also against walls in kitchen gardens not fully furnished with trees, – in short, in all bare spaces so often found between wall trees in old gardens. These methods of cultivating choice pears and the finer kinds of American apples are worthy of much more attention than they have hitherto received.

The method of cultivating the Prince Engelbert and one or two other kinds of plums as vertical single cordons, has been practised here for some few years ; it is original, highly worthy of attention, and may be made a profitable venture, not only for the amateur but for the market-gardener.

The management of those charming structures – ground vineries, is in this edition more fully gone into than before; in short, all the modes of culture

hitherto recommended have been revised and made

as perfect as practice can make them, for it must be recollected that all the modes of culture here recommended have been well tested, and no foreign practice recommended till found adapted to our wet English climate, the mean temperature of which is just about two degrees too low for the choice kinds of fruits to ripen without assistance.

September 1870.

2

SECTION 2

THE
  MINIATURE FRUIT GARDEN,
  PYRAMIDAL PEAR TREES ON THE QUINCE STOCK.

There is no description of fruit tree more interesting to cultivate in our gardens than the pyramid, – a name adopted from the French, the originators of this species of culture. The word conical, would, perhaps, convey a better idea of the shape of such trees ; but as pyramidal trees are now becoming familiar things in English gardens, it is scarcely worth while to attempt to give a new name to these very pretty garden trees.

For gardens with a moderately deep and fertile soil, pears budded on the quince stock will be found to make by far the most fruitful and quick-bearing trees; indeed, if prepared by one or two removals, their roots become a perfect mass of fibres, and their stems and branches full of blossom-buds. Trees of this description may be planted in the autumn, with a certainty of having a crop of fruit the first seasonafter planting, – always recollecting that a spring frost may destroy the blossoms unless the trees are protected. It must always be recollected that pears on quince stocks are strictly garden trees, and not adapted for orchards.

The most eligible season for planting pyramidal pear trees is during the months of November and December, but they may be planted even until the end of March; in planting so late, no fruit must be expected the first season. Still I ought to say here that I have frequently removed pear trees on the quince stock in March and April, just as the blossom- buds %vere bursting, and have had fine fruit the same season, particularly if sharp frosts occurred in May. The buds being retarded, the blossoms opened after the usual period and thus escaped. The experiment is quite worth trying in seasons when the buds swell very early.

If root-pruned pyramidal trees be planted, it will much assist them if about half the blossom-buds are thinned out with a penknife just before they open ; otherwise these root-pruned trees on the quince stock are so full of them that the tree receives a check in supporting such an abundance of bloom. About ten or fifteen fruit may be permitted to ripen the first season; the following season one to two dozen will be as many as the tree ought to be allowed to bring to perfection ; increasing the number as the tree increases in vigour, always remembering that a few full-sized and well-ripened pears are to be preferred to a greater number inferior in size and quality.

Fig. 1.

In the engraving (Fig. 1 on the preceding page) I have given a faithful portrait of a pyramidal tree of the Beurre de Capiaumont pear budded on the quince: this was taken in 1846 ; the tree was then about ten years old, and had been root-pruned three times. Nothing could be more interesting than this tree, only six feet high, laden with fruit of extraordinary beauty; for in my soil, pears on quince stocks produce fruit of much greater beauty and of finer flavour than those on pear stocks. I have, however, introduced the figure as much to show its imperfection as its beauty : it will be observed that its lower tiers of branches are not sufficiently developed ; this was owing to neglect when the tree was young – the tipper branches were suffered to grow too luxuriantly. Summer pinching in the youth of the tree is the only remedy for this defect, if it be not well furnished below ; and a severe remedy it is, for *all* the young shoots on the upper tiers, including: the leader, must be pinched closely in May and June, till the lower ones have made young shoots of a sufficient length to give uniformity to the tree. This requires much attention.

The quenouille, or tying-down system, is not practised in France at the present day; and, in truth, it does look very barbarous and unnatural. The trees trained in this manner in the Potagerie at Versailles are mostly on quince stocks ; they are from twenty to forty years old, and are very productive, but very ugly; all the shoots from the horizontal and depressed branches are cropped off, apparently, in July, as M. Puteau, the director, is, I believe, adverse tothe pinching system of M. Cappe. I have not for many years observed a single quenouille in Belgium ; all are pyramids, even in the gardens of the cottagers, and in general they are very beautiful and productive trees. In many cases, when on the pear stock, they are too luxuriant and require root-pruning; but this is not understood by Continental fruit-tree cultivators.

Pyramids and bushes are the trees best adapted for small gardens, and not standards such as are planted in orchards. To those conversant with such matters, I need only point to the very numerous instances of rich garden ground entirely ruined by being shaded by large spreading standard, or half standard unpruned fruit-trees. Now, by cultivating pyramidal pears on the quince – apples in the same form on the paradise stock – the cherry as pyramids and dwarf bushes on the Cerasus Mahaleb – and the plum as a pyramidal tree – scarcely any ground will be shaded, and more abundant crops and finer fruit will be obtained.

THE YOUNG PYRAMID.

If a young gardener intends to plant, and wishes to train up his trees so that they will become *quite* perfect in shape, he should select plants one year old from the bud or graft, with single upright stems; these will, of course, have good buds down to the junction of the graft with the stock. The first spring, a tree of this description should be headed down, so as to leave the stem about eighteen inches long. If the soil be rich, from five to six and seven shoots willbe produced ; one of these must be made the leader, and if not inclined to be quite perpendicular, it must be fastened to a stake. As soon, in summer, as the leading shoot is ten inches long, its end must be pinched off; and if it pushes forth two or more shoots, pinch off all but one to three leaves, leaving the topmost for a leader. The side shoots will, in most cases, assume a regular shape; if not they may be this first season tied to slight stakes to make them grow in the proper direction. This is best done by bringing down and fastening the end of each shoot to a slight stake, so that an open pyramid may be formed – for if it is too close and cypress-like, enough air is not admitted to the fruit. They may remain unpruned till the end of August, when each shoot must be shortened to within eight buds of the stem. This will leave the tree like the annexed figure (Fig. 2), and no pruning in winter will be required.

The second season the tree will make vigorous growth; the" side shoots which were topped last August, will each put forth three, four, or more shoots. In June, as soon as these have made five or six leaves, they must be pinched off to three leaves, and if these spurs put forth shoots, which they often do, every shoot must be pinched down to one or two leaves, *all but the leading shoot of each side branch;* this must be left on to exhaust the tree of its superabundant sap, till the end of August, unless the tree is being trained as a compact pyramid (see page 12). The perpendicular leader must be topped once or twice ; in short, as soon as it has grown ten inches, pinch off its top, and if it break into two or three shoots, pinch them all but the leader, as directed for the first season; in a few years most symmetrical trees may be formed.

When they have attained the height of six or eight feet, and are still in a vigorous state, it will be

Fig. 2.

necessary to commence root-pruning, to bring them into a fruitful state.

If some of the buds in the stem of a young tree prove dormant, so that part of it is bare and without a shoot where there should be one, a notch, half an inch wide, and nearly the same in depth, should be cat in the stem just *above* the dormant bud. If this

/ c

Pig. 3. be done in February, a young shoot will break out in the summer.1

I have thus far given directions for those who are inclined to rear their own pyramids. Time and attention are required, but the interest attached to well-trained pyramids, will amply repay the young cultivator.

THE MATURE PYRAMID.

The preceding figure (Fig. 3) is a pyramidal tree in its second and third year, and such as it ought to be in July before its leading side shoots and leading upright shoot are shortened. This, as I have said, is best done towards the end of August. The shortening

must be made at the marks ; all the side shoots

must be shortened in this manner, and the leading shoot; no further pruning will be required till the following summer. The spurs a, a, a, are the bases of the shoots that have been pinched in June ; these will, the following season, form fruit-bearing spurs. The best instrument for summer and autumnal pruning is a pair of hooked pruning scissors, called also ' rose nippers.'

As the summer pinching of pyramidal pears is the most interesting feature in their culture, and perhaps the most agreeable of all horticultural occupations, I nrnst endeavour to give plain instructions to carry it out.

The first season after the planting, about the middle or end of June, the side buds and brancheswill put forth young shoots; each will give from one to three or four. Select that which is most horizontal in its growth (it should be on the lower part of the branch, as the tree will then be more inclined to spread) for a leader to that branch, and pinch off all the others to three leaves (see Fig. 3, a, a, a). If these pinched shoots again push, suffer them to make three leaves, and then pinch them to two leaves ; but if the horizontal branch has a good leader, it will take off all the superfluous sap, and prevent the pinched spurs from breaking; the buds will only swell, and the following season they will be fruit spurs. The upper shoots of the tree, say to about two feet from its top, should be pinched a week before the lower shoots : this gives strength to those on the lower part of the tree.

1 Bare places in the stems of pyramids, and in the branches of espaliers or wall trees, may be budded towards the end of August with blossom buds taken from shoots two years old. This is a very interesting mode of furnishing a tree with fruit-bearing buds.

Fig. 4 is a side branch in June, with its shoots not yet pinched; Fig. 5 is a side branch with its shoot a, a, pinched in June ; 6 is the leader of the side branch, which should be pinched or cut off at the end of August to c.

In spring the perpendicular leader of the preceding year's growth will put forth numerous shoots, which must be pinched in June in the following manner: those nearest the base, leave six inches in length, gradually decreasing upwards, leaving those next the young leading shoot only two inches long. The leader of these ready-formed pyramids need not be shortened in summer, as directed for younger trees ; it may be suffered to grow till the horizontal leaders are shortened in August, and then left six or eight inches in length; but if the trees are to be kept to six or

t ' seven feet in height under root-pruning, this leading shoot may be shortened to two inches, or even cut close down to its base. For tall pyramids of ten, twelve, or fifteen feet, it may be left from eight to ten

Fig. 4. Fig. 5.

inches in length till the required height be attained; it may then be cut to within two inches of its base every season.

I ought here to remark that pear trees differ in their habits to an extraordinary degree : some makeshoots most robust and vigorous ; others under precisely the same treatment, are very delicate and slender. In the final shortening in August this must be attended to; those that are very vigorous must not have their shoots pruned so closely as those that are less so ; indeed, almost every variety will require some little modification in pruning, of which experience is by far the best teacher. It will; I think, suffice if I give the following directions for shortening the leaders of the side shoots, and the perpendicular leaders: – all those that are very robust, such as Beurre d'Amanlis, Vicar of Wink- field, Beurre Diel, &c., shorten to eight or ten inches, according to the vigour of the individual tree ; those of medium vigour, such as Louise Bonne of Jersey, Marie Louise, and Beurre d'Aremberg, to six inches ; those that are delicate and slender in their growth, like Winter Nelis, to four inches ; but I must repeat that regard must be had to the vigour of the tree. If the soil be rich, the trees vigorous, and not root- pruned, the shoots may be left the maximum length; if, on the contrary, they be root-pruned, and not inclined to vigorous growth, they must be pruned more closely.

COMPACT PYRAMIDS.

If pyramidal fruit trees, either of pears, apples, plums, or cherries, are biennially removed, or even thoroughly root-pruned, without actually removing them, summer pinching becomes the most simple of all operations. The cultivator has only to look over his trees twice a week during June, July, and August (penknife in hand), and cut or pinch in every shoot on the lateral or side branches that has made fourleaves or more, down to three full-sized leaves. It is just possible that the three buds belonging to these three leaves will put forth three young shoots: as soon as they have made their four or five leaves, they must be shortened to two, and so on, with *every young shoot* made during the summer, shortening the leading shoot also to three leaves. This method of close pinching represses the vigour of the tree to a great extent, and, in soils that are not very rich, trees under it will not require root-pruning. It is a most agreeable method of treating pyramidal trees, for no straggling shoots are seen, and in small neatly- kept gardens this is a great relief. The pinched shoots in these compact pyramids become too much crowded with blossom-spurs ; these should therefore be thinned in winter with a sharp pruning knife, removing at least one-third of them. It must be remembered that close pinching is applicable only to trees on the quince stock.

BOOT-PRUNING OF PYRAMIDAL PEAR TREES ON
QUINCE STOCKS.

Before entering on the subject of root-pruning of pear trees on quince stocks, I must premise that handsome and fertile pyramids, more particularly of some free-bearing varieties, may be reared without this annual or biennial operation. If the annual shoots of the tree are not more than eight or ten inches long, no root-pruning need be done. I have a large plantation of pear trees on the quince stock, which have been made very handsome and fertile pyramids, yet they have not been root-pruned, neither do I intend to root-prune them. But I wish

to impress upon my readers that my principal object is to make trees fit for small gardens, and to instruct those who are not blessed with a large garden, how to keep the trees perfectly under control: and this can best be done by *annual,* or at least biennial attention to their roots; for if a tree be suffered to grow three or more years, and then be root-pruned, it will receive a check if the spring be dry, and the crop of fruit for one season will be jeopardised. Therefore, those who are disinclined to the annual operation, and yet wish to confine the growth of their trees within limited bounds by root-pruning, say once in two years, should only operate upon half of their trees one season;1 they will thus have the remaining half in an unchecked bearing state; and those who have ample room and space may pinch their pyramids in summer, and suffer them to grow to a height of fifteen or twenty feet without pruning their roots. I have seen avenues of such trees in Belgium, really quite imposing. In rich soils, where the trees grow so freely as to make shoots eighteen inches in length in one season, they may be root- pruned annually with great advantage.

The following summary will, perhaps, convey my ideas respecting the management of pyramids andbushes when cultivated as garden trees. In small gardens with rich soils, either root-prune or remove all the trees annually early in November. In larger gardens perform the same operation biennially at the same season. For very large gardens with a dry good sub-soil, in which all kinds of fruit trees grow without any tendency to canker, and when large trees are desired, neither remove nor root-prune, but pinch the shoots in summer, thin them in winter when they become crowded, and thus make your trees symmetrical and fruitful.

1 In *The Journal of Horticulture* for 1862, page 531, Mr. Lee, of Clevedon, gives an account of his root-pruning practice, which he carries out extensively on some hundreds of trees. It appears to bo an alternate system of root-pruning, and may be done something after the following manner: – Open a semi-circular trench on one side of the tree, and prune all the roots that can be got at; the following season open a trench of the same shape on the opposite side of the tree (so as to complete the circle), and prune all that can be found there. By this simple method the tree is never checked seriously in its growth, yet enough to make it form abundance of blossom-buds.

Pyramidal pear trees on the quince stock, *where the fruit garden is small,* the soil rich, and when the real gardening artist feels pleasure in keeping them in a healthy and fruitful state by perfect control over the roots, should be annually operated upon as follows: – A trench should be dug round the tree, about eighteen inches from its stem, every autumn, just after the fruit is gathered, if the soil be sufficiently moist – if not it will be better to wait till the usual autumnal rains have fallen; the roots should then be carefully examined, and those inclined to be of perpendicular growth cut with the spade, which must be introduced quite under the tree to meet on all sides, so that no root can possibly escape amputation. All the horizontal roots should be shortened with a knife to within a circle of eighteen inches from the stem,1 and all brought as near to the surface as possible, filling in the trench with compost for the roots to rest on. The trench may then be

1 If they have not spread to this extent the first season, or even the second, they need not be pruned, but merely brought near to the surface and spread out.

filled with the compost (well-rotted dung and the mould from an old hot-bed, equal parts, will answer exceedingly well) ; the surface should then be covered with some half-rotted dung, and the roots left till the following autumn brings its annual care. It may be found that after a few years of root-pruning, the circumferential mass of fibres will have become too much crowded with small roots, in such cases thin out some of the roots, shortening them at nine inches or one foot from the stem. This will cause them to give out fibres, so that the entire circle of three feet or more round the tree will be full of fibrous roots near the surface, waiting with open mouths for the nourishment annually given to them by surface dressings and liquid manure.

Thus far for the gardener who does not mind extra trouble, – who, in short, feels real pleasure in every operation that tends" to make his trees perfect in frnitfulness and symmetry. But it is not every amateur gardener that can do this, nor is it always required in the south of England, except for small gardens and in rich moist soils, in which pear trees are inclined to grow too vigorously. But with our too often cool moist summers in the northern counties, annual root-pruning is quite necessary to make the trees produce well-ripened wood. In other cases, as I have before observed, shortening the shoots in summer, taking care to produce a handsome pyramidal form, and if they are inclined to grow vigorously, biennial root-pruning will be quite sufficient.

The following will be found a good selection of varieties for pyramidal trees on quince stocks. They may be planted in rows, five to six feet apart, or asquare may be allotted to them, giving each plant five or six feet, which will be found amply sufficient for root-pruned trees. Some few esteemed sorts of pears do not grow well on quince stocks, unless ' double-grafted '- – *i. e.,* some free-growing sort is budded on the quince, and after having been suffered to grow for one or two seasons, the sort not so free- growing is budded or grafted on it. For ten varieties, placed in the order of their ripening, the first ten may be selected, or if twenty, the second ten may with safety be recommended.1

1. Summer Doyenne' ..... July
2. Beurre' Giffard August
3. Beurre' de l'Assomption .... August
4. Bon Chretien (Williams') . . . September
5. Beurre' Superfin October
6. Louise Bonne of Jersey . . . . *m* § *e.* October
7. Doyenne du Comice November
8. Beurre' d'Aremberg .... December
9. Josephine de Malines .... March
10. Bergamotte d'Esperen2. . . . April and May
1. Madame Treyve August
2. Souvenir du Congres .... September
3. Baronne de Mello October
4. Beurre' Hardy October
5. Doyenne' Gris *e.* October
6. Conseiller de la Cour . . . . *b.* November
7. Winter Nelis December
8. Beurre' d'Anjou *e.* December
9. Easter Beurre' ...... January
10. Doyenne d'Alencon .... March to May

(In the above list, varieties marked thus may be chosen by those who require only a few trees.)3

1 All the varieties recommended for pyramids may also be planted as espaliers to train to rails in the usual mode.

" This is a most abundant bearer. A pyramid in the garden of Thomas White, Esq., which was root-pruned in the autumn of 1858, bore two bushels in 1859.

3 A very good light permanent label for pyramidal and other fruit trees is a small piece of zinc, painted with white-lead paint, and written on while moist with a strong black-lead pencil. It should be suspended from a side branch of the tree (not the stem) by a piece of stout copper wire.

The above succeed on the quince, and form excellent pyramids.

PYRAMIDS FOR MARKET GARDENS.

The culture of pears for sale, in favourable soils and sites, has been found very profitable ; it is a pleasant method of adding to one's income through an occupation of pleasure. This amateur market gardening is at present in its infancy : it will go on and increase to an extent not yet thought of; I will therefore give a few hints as to culture.

First, a good climate must be selected somewhere south of the Trent, the site sheltered from the north and east and north-west by hedges, evergreens, or walls ; also a favourable soil, which, however, by care and culture, may be made of secondary importance ; a loam eighteen or twenty inches deep, on a dry stony sub-soil, is perhaps the most favourable, but a clayey loam resting on clay or on sand will do very well. If required, draining must be practised, so that clays, loams, or sands must be dry.

When a rich deep fertile soil is planted there will be nothing required but opening the holes and planting the trees ; but if the soil be shallow, say less than twelve inches of staple, it should be stirred to a depth of twenty inches, leaving the stirred subsoil *in situ*. The soil is thus far prepared for planting, which will be best done in November, December, or February. The trees should be planted six feet apart row from row, and the same, tree from tree in the row. After the trees are planted, the soil within a circle of three feet round the stem of each treeshould be trodden firmly ; a small portion (the tenth of a barrowful) of litter or manure placed round each tree, or if the soil is rich this may be omitted, and the work is done. For some four or five years the centre ot the space between the rows of trees may be cropped with light vegetable crops, such as onions, &c.; this cultivated space must be confined to a width of two feet, the remaining space next the trees must not be touched with anything but the hoe to kill the weeds, and when the intermediate cropping has covered the entire surface of the ground, it must remain firm, the only culture besides the hoe being an occasional surface dressing of manure. This system of hard soil and occasional snrface manuring is the *summum tionum,* the last step towards perfect market garden fruit culture, except gooseberries, currants, and raspberries, which require other treatment. The quantity of manure required for a surface dressing is five bushels to twenty-five square yards.

The rough and ready pruning now practised on my market garden pyramidal pears is as follows : – Towards the middle (the end, if the season be late) of June all the young shoots are shortened to half their length with pruning scissors, and towards the end of August the trees are again gone over, and two-thirds of all the young shoots, some of which are made from the shoots pruned in June, are cut off; this is all the pruning required, unless the amateur market gardener pleases to amuse himself in winter by removing a crowded shoot or shortening a spur, making good use of his eyes. The employment is cheering and the profit is agreeable. The varietiesbest adapted for this mode of pear culture are few, as there are but *few* sorts popular in the markets. Our first and best is Louise Bonne, and then follow Beurre d'Amanlis, Madame Treyve, – a charming August pear, – Williams' Bon Chretien, Beurre Clairgeau, – this is best double-grafted on Vicar of Winkfield, – Beurre Hardy, Doyenne da Comice, Souvenir du Congres, Beurre Superfin, Marie Louise (d'Ucele), and Beurre de l'Assomption ; these are all large and excellent fruit, and come in season, while pears are popular. Beurre Clairgeau is the last of our market pears, ripening at the end of November and in December. This is one of the most beautiful pears known, and in a warm season and in a warm soil, is melting and very good. I have a plantation of pyramids about five hundred in number, all double- grafted on the Vicar of Winkfield, which were trees on the quince, two years old from the bud when double-grafted; every tree is a picture of health and fertility. I should add that this plantation is of trees six feet apart row from row, and three feet apart tree from tree ; and I may also state that this sort in some soils grows freely on the quince without being double-grafted.

I have named ten sorts fitted for amateur market gardening; if any one is inclined to select two sorts, I should say take Louise Bonne and Beurre Clairgeau.

ORNAMENTAL PYRAMIDAL PEAR TREES ON QUINCE STOCKS.

There are some few varieties of pears, the trees of which may be made highly ornamental even on a well-dressed lawn, as they grow freely and formnaturally beautiful cypress-like trees; at the same time their fruit is of first-rate quality. Such are Summer Beurre d'Aremberg, Baronne de Mello, Duchesse d'Angouleme, Urbaniste, Alexandra Lam- bre, Beurre Hardy, White Doyenne, Grey Doyenne, Louise Bonne of Jersey, Passe Colmar, Zephirin Gre- goire, Beurre de l'Assomption, Souvenir du Congres, Beurre Leon le Clerc, Delices d'Hardenpont, Doyenne du Cornice, Bergamotte d'Esperen, and some others.

PEAR TREES AS BUSHES ON THE QUINCE STOCK.

It is only very recently that this mode of cultivating pear trees has struck me as being eligible, from having observed the fruit of some of the large heavy varieties, such as Beurre Diel and Beurre d'Amanlis, so liable to be blown off pyramids by even moderate autumnal gales. The trees also of these and several. other fine sorts of pears are difficult to train in the pyramidal form; they are diffuse in their growth, and, with summer pinching, soon form nice prolific bushes, of which the following figure (Fig. 6), from nature, will give some idea. This summer pinching is quite necessary in bush culture, and is performed by pinching off the end of every shoot as soon as it has made four or five leaves, to three full-sized ones; the buds at the base of these leaves will each put forth a shoot, which should be pinched to two leaves, these will again put forth young shoots, which should be pinched to one leaf; this is the third summer pinching, and probably in August a fourth may be required, as pear trees in a good soil and good climate are very persistent in their growth daring the warm weather of summer. All the shoots after the second pinching should be pinched to one leaf. By repetition of this summer pinching the trees become crowded with spurs; these will require to be thinned in winter, removing at least one-half of them, for there is a

Kg. 6.

great tendency in pear trees that are grafted on qnince stocks to produce too many blossom-buds. The biennial removal described below is also necessary, unless in very large gardens where large spreading trees are wished for. Although the taking up and replanting a tree may seem formidable work, it is not so, for the roots, from being frequently removed,

become so fibrous near the surface, that an active man can lift and replant one hundred trees in a day. I need scarcely add that if root-pruning, as described in page 13, be preferred to removal it may be practised.

These bushes are admirably adapted for gardens exposed to winds, and if removed biennially they may be grown in the smallest of gardens with great advantage. This biennial removal or lifting, should be performed as follows : – A trench should be opened round the tree the width of a spade, and from twelve to fifteen inches deep ; the tree should then be raised with its ball of earth attached to its root intact. If the soil be light and rich, and the tree inclined to grow vigorously, making annual shoots of more than one foot in length, it may be replanted without any fresh compost; but if, on the contrary, the soil be poor, and the tree stunted in its growth, the following materials may be used : – In low situations near brooks and rivers, a black moor-earth is generally found; this unprepared is unfit for horticultural purposes, but if dug out and laid in a ridge, and one-eighth part of unslacked lime be spread over it, turning it immediately and mixing the lime with it, it will become in the course of five or six weeks an excellent compost for pear trees. It is a good practice to add half-a- busbel of burnt earth, or the same quantity of sand, to a barrowful of thismoor earth. Leaf mould (or rotten manure), loam, and sand, equal parts, form also an excellent compost; in planting, one wheel- barrowful to a tree will be enough. In Londonsuburban gardens, for which these trees are peculiarly adapted, no compost need be given to the trees in replanting, for the soil iu them is generally rich. These bush trees offer two very great advantages, they are easily protected from spring frosts when in blossom by covering them with tiffany, and they may be planted from three to five feet apart with great facility, so as to be eligible for very small gardens.

In large gardens in situations exposed to the wind, large bushes may be desirable. In such cases the leading shoots on each branch may bo pinched, as recommended for pyramids (page 9) ; but instead of pinching them to three leaves, they may be suffered to make ten leaves, and then be pinched, leaving seven. The trees will if treated in this manner, soon become large, compact, and fruitful.

The following varieties are well adapted for bush culture, as they are spreading in their growth and difficult to form into compact pyramids, although they may be made into spreading and prolific conical trees. It ought, however, to be mentioned that those sorts, such as Louise Bonne of Jersey, which form handsome pyramids, make very pretty compact bushes by cutting out the central branch to within three feet of the ground, so that pyramids may be easily formed into bushes. I may add that these bush pears produce the very finest fruit, from their being so near the heat and moisture-giving surface of the earth.

In situations near the sea-coast, exposed to sea breezes, small fruit gardens may be formed by enclosing a square piece of ground with a beech hedge or wooden fence, and planting it with bush trees. A piece of ground 500 square feet will be large enough to cultivate 30 trees at 4 feet apart in it, or 25 trees at 5 feet apart. Many a sea-side cottage may thus have its fruit-garden.

LIST OF PEARS ADAPTED FOR BUSH CULTURE.

Autumn Nelis. ...... October
Beurre " d'Amanlis September
Beurre" de Caen October
Bern-re" de Ranee . . . . . March
Beurre" Diel .... December
Beurre" Giffard....... August
Catillac (for baking) December
Conseiller de la Cour November
Doyenne1 Boussoch October
Doyenne" du Cornice ..... November
Jalousie de Fontenay ..... August
Jargonelle August
Josephine de Malines March
Le"on le Clerc de Laval (for baking) . . March
Marie Louise .... October
Ollivier de Serres February
Passe Crassane February
Prince of Wales (Huyshe) .... December
Victoria (Huyshe) November
Winter Nelis December
Zephirin Gregoire January

PROTECTORS FOR PYRAMIDAL AND BUSH PEAR TREES.

The weather in spring is often cold and ungenial for the blossoms of pear and other fruit trees; in such seasons pyramids should be protected. This is best done by fixing four stout stakes round a tree; these should be a little taller than the tree andv then be sawn off level. A square piece of calico, or any cheap canvas, should then be nailed on the top of the stakes to form the roof, the like material brought round the sides, and fastened to the stakes by small nails or tacks, from within eighteen inches of the gi Jtmd to within eight inches of the top, thus leaving a space between the top and side covering for free ventilation, as the air, when heated by the sun, will rush out of the aperture at top in a continual stream. These flat-roofed square tents will generally iiu. *Ire* a crop of fruit.

Pea-sticks – *i. e.,* stakes with the small brush-wood on them – stuck round each tree, and spruce or other fir branches, where these can be procured, are also good protectors. For bush trees hay is a capital

THR TIFFANY HOUSR PROTECTOR.

Section of Tiffany House.

protector, particularly from those still hoar frosts which are generally so destructive; it should be strewed lightly over them when they are just commencing to bloom. If some brushwood sticks are placed round the bush so as to lean over it, the hay will adhere to the spray, and remain undisturbed by the wind. Tiffany – double tiffany is to be preferred as it is warmer – may be used to throw over pear bushes; it is so light that it does no injury to the tender blossoms; it should be taken off on sunny days. There is, perhaps, no better protector than

old or new netting ; if woollen, all the better. This should be thrown over the trees two or three times thick, and suffered to remain on till the fruit is safe from frosts – *i. e.* till the end of May.

Houses built with stakes or slight timber, and the roofs and sides covered with tiffany, have very recently been introduced and found efficient in protecting half-hardy plants from severe frost.

I now propose to erect temporary houses of the same material to protect dwarf and pyramidal fruit trees while they are in bloom, and I have no doubt but that they will lead to a new era in fruit gardening among amateurs, offering as they do a very cheap method of protection. A border or bed of fruit trees may be eight feet wide and planted with three rows of bush fruit trees as shown in the above section, one row in the centre, and the other rows three feet from it, and the trees three feet apart in the rows, thus occupying six feet of the bed.

A tiffany-house to cover the trees in a bed of the above width may be eight feet wide, three feet high at the sides, and five feet high in the centre.

The roof of tiffany should be fastened to the rafters with shreds three or four times double-, so as to make a thick pad, and either nailed on with short nails or fastened with screws, so that it may be easily taken to pieces annually the first week in June, for till then we are not safe from spring frosts. The tiffany house should be placed over the trees the first -week in March, unless the season be unusually early, when the middle of February would be better. The sides should be loose, and be turned up night andday in mild weather while the trees are in bloom; but in cold, sharp, windy weather, in the blossoming season they should be kept down, and fastened to the upright stakes by tying or otherwise.

A tiffany house twenty-four feet long and eight feet wide will thus shelter twenty-four trees, either bushes or pyramids; if for the latter, the sides of the house should be four feet, and its centre seven to eight feet in height. If it be thought desirable to keep the trees in a comparatively small space, they may be removed biennially in October. If larger trees are desired, the house may be enlarged as the trees grow. A tiffany-house may be from one to 500 feet in length, and twenty in width if desirable, for there are no particular limits to its extent, only the effects of a 'March wind' mnst be thought about when lofty and extensive houses are put up. As measures of economy the timber and tiffany should be placed in a dry place when removed and the rafters fastened to the plate and ridge board with screws. A tiffany-house thus treated ' kindly and gently,' will last for several years ; and in places where the climate is sufficiently warm to ripen apricots, plums, pears, cherries, and even early peaches, in the open air, they will, I have no doubt, be extensively employed.

PEAR TREES ON THE QUINCE STOCK, TRAINED AS COItDONS.

The French gardeners employ the term cordon for the branch of a fruit tree on which the shoots have been pinched in, so as to form a succession ofblossom-buds. The term as used by them, is expressive, and lately an interesting work has been published by the Rev. T. C. Brehaut, of Guernsey, on this mode of training, under the title of ' Cordon Training of Fruit Trees.' It is simply the pinching off the ends of the shoots on a branch so as to make them form blossom-buds, and fruit trees under this mode are planted in an oblique position on walls. With pear trees on the quince stock the five-branched vertical cordon will be found a very convenient mode of training, for which see Fig. 7. To carry out this mode of training, in April, 1849, I planted one of each of some new and esteemed pears on quince stocks against a boarded fence, so that they would quickly come into bearing. The usual method of horizontal training I found would take up too much space, and I could not find room for half the number of trees I wished to plant. In this strait an old idea came to my assistance – that of cutting pyramidal trees flat, and planting them against walls; and then a modification of the idea came to hand, viz., to plant horizontal espaliers, and to make them perpendicular. In the next page is a figure of one of my five-branched vertical cordon pear trees (Fig. 7).

The shoots, *a, a,* should be eight inches from the central shoot, and those marked 6, 6, the same distance from those marked *a, a.* This tree, with five branches, will thus occupy thirty-two inches – say three feet of wall room; a tree with seven branches will require four feet, but as some spaceought to be allowed for the spurs on the outside branches, say five feet. If the wall be of a moderate height, eight feet for instance, a tree with seven branches will produce quite enough fruit of one sort. This method offers a strong contrast to espaliers on pear stocks, planted in the usual manner twenty-four feet apart and trained horizontally;

A FIVE-BRANCHED VERTICAL CORDON PEAK TREE.

6 a 06

Fig. 7.

nearly five trees for one will give so many additional chances to the pear cultivator ; the single tree may fail, or its fruit may become imperfect, owing to an adverse season ; but out of his five trees, he will, in every season, stand a good chance of having *some* good pears. A few words will suffice for their management; summer pinching of the shoots to three leaves all through the summer, as recommended for pyramids (p. 9), and root-pruning or biennial removal, these operations, – like Dr. Sangrado's bleeding and warm water – will do all.

Five or seven-branched vertical cordon trees, not only of pears but of cherries on the Mahaleb stock, of plums, and of American apples on the Paradise stock, and peach and apricot trees may be planted against walls in gardens, if of a moderate height, to great advantage. As so much variety may be had in a small space, let the reader imagine himself to have a brick wall with a southern aspect, 20 feet long and 8 or 10 feet high. According to old practice this would afford space for *one* tree, but with branched vertical cordon training, I repeat *five* trees may be cultivated, and thus give five chances to one.

If this kind of tree on the quince stock cannot be procured, those that are trained horizontally, with five or seven branches, may be planted against the wall or fence destined for them; and their young shoots a, a, and *b,* 6, in Fig. 7, be made to curve gently till they are perpendicular ; the young shoots of pear trees are very pliable, and will easily bend to the required shape. The lower part of each shoot in such cases must be fastened to the wall with shreds and nails, in the usual way, and the remaining part brought round to an upright position. If they are more than two feet in length, each of these shoots must then be shortened to it. These shortened branches will, in May, each put forth two or three shoots. As soon as they have made five or six leaves pinch all but one on each branch to three leaves, leaving the topmost one to each shoot, *a, a,* and *b, b,* as above, also to the leader. You will thns, if your tree be five-branched, have five young leading shoots. As soon in June as they have attained to eight inches in length, pinch off the end of each, and when they break into two or three shoots as before, pinch so as to leave the spurs with three leaves, and the leading shoot unpinched to each branch. This may be repeated if the soil be rich, two three or four times in the summer. Tour tree will soon reach the top of the wall and every bud in the five branches will be perfect, either a blossom-bud or one in embryo. When every branch has reached the top of the wall, commence root-pruning in autumn, unless the trees have ceased to grow vigorously and are bearing well – if so, leave their roots untouched; the directions for root-pruning are given in treating of pyramidal trees (p. 13). These may be followed exactly ; and, if so, the trees will be kept in a stationary bearing state. It must be recollected that the spurs on the branches will often put forth shoots even while bearing fruit; these must be pinched into three leaves.

I may as well hint to the reader, that, if larger trees are wished for, so as to give more fruit of each sort, trees with nine upright branches may be planted seven feet apart, or trees with eleven upright branches, nine feet apart. Trees, however, can seldom be purchased with shoots so numerous ; young trees must therefore be planted, and cut back annually for two or three years, till the proper number of perpendicular shoots are supplied. It may happen that trained trees with five or seven branches cannot be procured, perhaps trees with only three shoots, two horizontal and one leading shoot; in snch cases they must be cut back, leaving five buds to each shoot, and the young shoots in June trained as required.

Pig. 8.

Pyramidal trees cut flat on the side to be placed next the wall, and planted against walls or fences will give almost a certain crop. Their shoots must be pinched, and trained so as to form a handsome semi-pyramidal tree, which when it has reached thetop of the wall, must be subjected to biennial root- pruning ; but this will only be necessary if the tree is too vigorous, so as to keep it in a stationary fruitful state. Annexed I give a figure (Fig. 8) of a young pyramid planted against a south-east fence.

It will, I trust, be seen how economical of space are these methods of training pears to walls; and I know of nothing in fruit culture more interesting than a wall of upright five-branched cordons or of pyramids full of fruit. Let us only consider that a wall 100 feet long will accommodate *five* trees on the pear stock, trained in the usual horizontal mode ; the same wall will give ' ample room and verge enough,' to *twenty-five* trees on the quince stock, trained perpendicularly ; if their young shoots (all but the leaders) are pinched into three leaves all the summer, no root-pruning will be needed. They are also invaluable for planting against walls between old trees where there are bare spaces, as is so often the case, for they soon fill up such vacancies, and bear abundance of fine fruit. A selection of varieties for wall trees will not here be out of place.

UPRIGHT TRAINED TREES ON QUINCE STOCKS.

FOB SOUTH OB SOUTH-WRST WALLS.

Crassane Glou Morceau

Summer Doyenne"1

Chaumontel

Passe Colmar

Hardy

Van Miis (Leon le Clerc)

Gansel's Bergamot2

1 This will ripen on walls towards the end of June, quickly followed by Citron des Cannes.

3 It is not generally known that this fine variety, proverbially a shy bearer, becomes, when double-grafted on the quince stock, one of the most abundant bearers.

UPRIGHT TRAINED TREES, Ac. – *continued.*

FOR WEST OR NORTH-WEST WALLS.

Beurrt Diel

Beurre' d'Amanlis

Beurre' de Ranee

Passe Crassane

Beurre' Superfin

Marie Louise

Louise Bonne of Jersey

Josephine de Malines

FOR EAST OR SOUTH-EAST WALLS.

Beurre' Easter

Beurre' d'Aremberg

Bergamotte d'Esperen

Winter Xelis

Beurre' de l'Assomption

Doyenne' d'Alencon

Beurre de Caen

Conseiller de la Cour

Beurre' d'Anjon

Souvenir du Congres

The above varieties grafted on pear stocks are equally adapted for their several aspects. In shallow, gravelly, or chalky soils, pears on pear stocks are to be preferred for walls.

It is almost useless to plant dessert pears against north or north-east walls, as the fruit, unless in very warm seasons, is generally deficient in flavour. The only varieties that offer the least chance 'of success, and that only in a warm climate with a dry soil, are Marie Louise, Jargonelle, Louise Bonne of Jersey, Beurre Superfin. It is far better to plant against such aspects baking or stewing pears, such as Catillac, Bellissime d'Hiver, and Leon le Clerc de Laval; the Vicar of Winkfield is also a good north wall pear – it bears well and stews well. In the north the finer sorts of pears must be cultivated on south walls.

In recommending pears on quince stocks as pyramidal trees for cold soils and situations, even in the far north, I may appear theoretical; but from my own experience in some very cold and clayey soils in this neighbourhood, I feel sanguine as to the result, for I have observed in my frequent visits to the pear gardens of Trance that many sorts are often *too ripe.*now, this is just the tendency we require. In our cold and moist climate, most certainly, pears will not get *too ripe,* more especially in the north of England and Scotland. Some years since I received a letter from a correspondent living in a hilly part of Derbyshire, from which I have an extract : – ' I have tried Beurre Diel, Beurre de Capiaumont, Marie Louise, and Williams' Bon Chretien, on pear stocks, all of which bear well as standards, but their fruit does not come to perfsction, always remaining quite hard till they decay at the core. I have placed the fruit in a hot-houe, but hae never succeeded in ripening them. Williarc-s' Bon Chretien we can only use for stew 'ng.' This seems to show that cold hilly situations are not favourable to the cultivation of pears as standards. I have recommended some pears on quince stocks, and have heard of a favourable result.

CORDON PEARS ON TRELLISES UNDER GLASS.

Some few years since a very ingenious method of growing peaches and nectarines on trellises, over which were placed moveable glass lights, was invented by Mr. Bellenden Ker. In warm and sheltered ga.'dens this mode of culture answers very well for peaches, but in cool climates there is not day-heat enough stored up, as in houses, to act upon the fruit. Cheap orchard-houses are, therefore, to be preferred to these cheap trellises for the above kinds of fruits, unless the garden be small and much sheltered.

Soon after I had built my trellis for peaches, it occurred to me that the system applied to pear culturewould do well, and so I built the trellis 60 feet long and seven feet wide ; on this I planted upright five- branched cordon pear trees on quince stocks (see Fig. 7). Fig.9 is a section of this trellis, and Fig. 10 is a

Fig. 9.

front view of a pear tree trained to it in the upright method. My trellis was planted ten years ago, and

Fig. 10.

has now on it twenty fine trees, about fourteen years old, and in fall be. ring. They were planted th'-ee feet apart, as it was my first experiment, and are now a little crowded ; four feet apart will be found the proper diVTice. I have never seen anything more interesting in fidit ciiare than this trellis covered with pears, for, owing to its being near the ground, te radiation of heat gives the .-uit a size and beauty rarely seen even on walls.

The lights should remain over the trees till tte beginning of July, and then be removed, sufferir 5 the fruit to ripen felly exposed to the sun and air. It seems that the glass over the fruit in its youngstate serves to develop its growth in a remarkable manner, for rarely is a spot seen on pears grown on. these trellises; they have a clear, beautiful appearance, much like those grown in the warmer parts of France. I ought to add that, in cool climates, such as the North of England and Scotland, the lights may be suffered to remain over the trees till the beginning or middle of August. This will hasten the ripening of the fruit, but it should be exposed to the air in early autumn for some weeks before it is gathered (unless the climate be particularly cold and stormy), or it may suffer in flavour. Pears ripened under glass are apt to suffer in this respect. I have, however, very recently received the following communication from a very clever fruit-cultivator living in Ireland : – ' Let no one persuade you that pears grown in a well-ventilated orchard-house are not equal to those outside; I can give strong evidence to the contrary. In my house there was a small Louise Bonne on the quince stock, in an 11-inch pot; it bore 23 splendid pears, as far superior to the same fruit grown in the open air as it was possible to be. They were not, I admit, high-coloured, but they attained a richness and flavour that I thought Louise Bonne did not possess.'

The pear trellis of which the section and front view (Figs. 9 and 10) will give a correct idea, is of the most simple description. A row of larch or oak posts must be driven into the ground 6 feet apart, and another row in front; on these should be nailed plates, 3 inches by 2, and then bars, 3 inches by 1, placed flatwise, from front plates to back 3 feetapart; across these common tiling laths should be nailed, six inches asunder. This will form the trellis as seen in Fig. 9. The supports for the lights are formed in the same manner, by a row of posts at the back, and the same for the front, on which are nailed plates of the same dimensions as those for the trellis ; a cross-piece should be nailed to front and back plate at each end, to keep the supports for the lights from giving away. The structure with the lights, when resting on the back and front plates, has exactly the appearance of a large garden frame without back, front, or ends. Under the lights the trellis is formed with a sharp slope upwards to the back; for unless the front of the trellis is within six inches of the ground, it will be difficult to bend the trees to the required position. By this simple contrivance, pears (and even peaches and nectarines in warm gardens) may be grown in any corner of the garden, with a south or south-western exposure – for it is scarcely necessary to add that the lights should slope to the south or south-west, so as to have all the sun-heat possible.

The most eligible dimensions for a trellis I find from experience to be as follows :
–

Glasi Lights.
Eight feet long, three feet wide.
Height from ground at back, three feet six inches.
Height from ground at front, one foot six inches.
Trellis.
Height from ground at back, two feet six inches.
Height from ground at front, six inches.
Distancefrom glass lights, one foot.

The front border should be raised to a level with tne front of trellis ; this will leave twelve inches betweenthe front ends of the lights, and the surface of the front border, which will be quite enough for ventilation ; indeed the draught in windy weather is inclined to be too sharp. I find, therefore, furze, or other evergreen branches, placed along the front between the glass and the border, and a mat nailed at the back, excellent checks to excessive ventilation in cold frosty weather. They may remain there till the beginning or end of June ; the latter if the weather be cold and stormy. The lights are fastened to the plate, back and front, by a hook and eye ; they are thus easily removed to prune the trees and gather the fruit.

I was induced, as I thought, to improve upon Mr. Ker's plan, by having my first trellis within eight inches of the glass, – for I calculated, the nearer the glass the better the chance of success in early ripening, but I suffered for my innovation. My peach trees were planted in March, 1848 ; they made during the summer, with the lights constantly on, beautifully matured shoots, and in March and April, 1850, were gay with blossom. The winds were cold, the nights frosty ; but owing to the extreme ventilation, which kept every bud and shoot dry under the glass, not a blossom was injured by the sharp winds, and the trees were covered with fruit. On the fatal 3rd of May, however, in 1850, a still hoar frost – the thermometer down to 23 – destroyed all my hopes, for, owing to the trees being too near the glass, every fruit was blackened and destroyed: a single mat would have saved them, but I was not at home, and my pet trees were forgotten. Do not, therefore, have the trellis nearer the glass than twelve inches.

3

SECTION 3

It will be seen that I employ smaller lights, which vre easily removable for purposes of culture, and a smaller trellis than that described by Mr. Ker in the seventh edition of this work. I find from experience this smaller edition of the Kerian trellis much to be recommended for small gardens. The Kerian trellis has answered so well for pear cultivation that I have been induced to amplify the original idea ; and in the Appendix is given a diagram of a trellis I have recently made.

HORIZONTAL CORDON PEAR TREES ON DWARF WALLS.

Having had occasion within these two years to erect a large number of four-inch brick walls on which to train young peach trees, I have been much struck with their eligibility for pear trees on quince stocks. A very large number of trees may be cultivated in this manner on" a small piece of ground.

My walls have a nine-inch foundation of four courses of brickwork in the ground, and they are carried up to four feet above the surface (it is scarcely safe to build them of great height), with nine-inch piers fifteen feet apart. The coping for them is made of boiling coal-tar mixed with lime and sand to the consistence of mortar, which is placed on the top of the wall thus *jj* so as to carry off the water. This is a most cheap and efficacious covering – it can scarcely be called a coping, as it does not project over the edge of the wall. A coping of Portland cement is even better, as it holds the wall together.

The best description of brick for these light walls is the patent perforated brick, but common stockbricks will do. The very best lime should be used (I have found the grey Dorking lime excellent), but any kind of lime made from limestone will answer well; that made from chalk in this country is not strong enough. Their cost, as I learn from my bricklayer, is about six shillings a yard in length; thus a wall of the above height, twenty yards long, should cost six pounds. In places where bricks are cheap they may be built for less ; if they are dear and at a distance, their carriage will add to the expense. My walls are six feet apart, and stand endwise, N. E. and S. W.; so that one side of each wall has a S. E. aspect, the other a N. W.; on the former may be grown the late-keeping pears, on the latter the earlier sorts, that ripen from October till the end of November. We thus have one excellent aspect – the S. E.; and one tolerably good – the N. W.; so that no wall space is lost.

The pear trees for these dwarf walls should be grafted on quince stocks, trained horizontally, pruned by summer pinching as directed for five-branched vertical cordons (p. 28). They may be planted five feet apart at first, and when their branches meet they should be interlaced, as in Fig. 11, and if necessary – *i. e.,* if the shoots be long enough – they may be trained over the stems, so that the wall is completely furnished with bearing branches. At the end of five or six years every alternate tree may be removed, leaving the permanent trees ten feet apart. I advise planting thus thickly, because I know from experience that the temporary trees will fill the walls, will bear a good quantity of fruit, and look moresatisfactory than if they are planted thinly. When removed they may be planted out for espaliers, or fresh walls built for them. I have some trees that have been planted six years ; but I find that, owing to the soil not being rich, they have not grown rapidly, and need not yet be removed, as their branches only just cover all the fence to which they are trained.

If, owing to the soil being rich, the trees are inclined to grow vigorously and not bear, they should

Fig. 11.

be lifted biennially, or root-pruned; but pears on quince stocks will be sure to bear abundantly.

These dwarf walls when covered with well trained trees, have a neat and charming effect, and the trees may be easily protected by sticking branches of evergreens in the ground and letting them rest against the wall, or by wooden shutters, placed on the ground at an angle, so as to *resi* against the wall; but I intend to be more luxurious, and to have cheap glass lights, in lieu of shutters, placed against the walls, and suffered to remain, so as to cover the trees till the fruit is fully formed, or till the first week in June, when all fear of damage from frost is over.

Where two or more walls are built, or a square piece of ground devoted to them, a cross wall or walls should be built at the north-east end, to prevent the sharp current of wind from the northeast, which would blow up the intervals between the walls with great violence. It is surprising what a quantity of fruit may be grown on a small space of ground with the aid of these walls ! Peaches, nectarines, and apricots may be grown on the S. E. aspect, but the trees must be kept in check by biennial removal. I have at this moment more than two thousand yards in length of them, and I intend to add to them annually, so convinced am I of their economy and utility. They seem to me more particularly suited to suburban, or what are commonly called cockney gardens. How pleasant to be able to have a brick wall twenty yards long for six pounds, or ten yards long for three pounds; and how delightful to be able to grow one's own ' wall fruit!' On a wall ten yards long, five peach and nectarine trees may be trained, and many dozens of fruit produced annually. These dwarf walls for the cultivation of peaches, nectarines, and apricots must, however, differ from those for pear trees, and be built so as to give a south or south-west aspect for the front, a north or northeast for the back. The latter may be planted with Morello cherries. To carry out the cultivation of the above-mentioned trees on dwarf walls, it is absolutelynecessary to take them up biennially in November, and replant them in the same place.1 They will not require any compost to then- roots, for peach, nectarine, and apricot trees are generally by far too vigorous in their growth. In some of the London suburban gardens the soil is so rich, that annual removal, *particularly with apricots,* may be found to be quite necessary. In country gardens where the soil is poor, a dressing of manure on the surface over the roots two inches deep will be of service. The peach trees on my experimental wall are removed biennially. The soil is not rich, yet they are almost too vigorous ; they bear fine fruit and give good crops.

A matter of great consequence in peach tree culture on walls is to keep the surface of the soil solid; if, therefore, the trees grow too vigorously, so as to require removal, say in October, the soil, after the tree is planted, should, after becoming dry, be rammed with a wooden rammer, so as to be as solid as a common garden path. In spring this hard surface should be covered with a slight coat of thoroughly decayed manure, which will be all the culture required.

ESPALIER PEARS ON QUINCE STOCKS.

Pears on the quince may be cultivated as horizontal espaliers or cordons by the sides of walks, or trained to lofty walls with much advantage, as less space is required. Horizontal espaliers or wall trees onthe pear stock, trained to walls of tlie usual height, *i. e.,* from ten to twelve feet, require to be planted twenty feet apart, while those on the quince may be planted only ten feet apart; this, in a small garden, will allow of much greater variety of sorts to supply the table at different seasons. With these the same high culture, if perfection be wished for, must be followed; the trees carefully planted, Bo that the junction'of the graft with the stock is even with the surface of the mould formed as directed for pyramids. The pruning of wall pear trees has always been a subject of controversy with gardeners, as they are inclined to grow too vigorously. If it be thought desirable to have trees of large growth, so as to cover a high wall, and yet be highly fertile, it is much better to root-prune than to prune the branches. With such trees it need not be done so severely : biennial root-pruning will be quite sufficient, commencing at eighteen inches from the wall after the tree has had two seasons' growth, cutting *off* the ends of all the roots at that distance from the wall, and increasing it by six inches at every biennial pruning, till a distance of six feet from the wall is reached. When this is the case the roots must be confined to the border of that width by digging a trench biennially, and cutting off all the ends of the roots at that distance from the wall.

1 It is a prudent practice in all cases of biennial removal to remove half the number of trees in alternate years, for in dry seasons those recently removed may be too much checked in their growJh to tear a crop "of fruit the first season after removal.

I may, perhaps, make this more plain by saying that a tree planted in November, 1860, should have its roots shortened eighteen inches in November, 1862 ; to twenty-four inches in 1864; to thirty inches in 186(f; to three feet in 1868; and so on, leaving six

inches biennially till, say, a distance of six feet from the wall is reached in 1880. This border, six feet wide, will then be full of fibrous roots.1 It should ever be dug or cropped, but annually have a surface dressing of manure, about two inches in thickness ; and, as I have before said, have a trench dug biennially eighteen inches deep, six feet from the wall, and the end of every protruding root cut off. If this method be followed, summer pinching to three leaves the first time, and to one leaf afterwards, of the spurs on all the leading branches, may be practised, and scarcely any winter pruning will be required.

In forming borders for wall pear trees on quince stocks biennially root-pruned, the soil should be well stirred with the fork to a depth of eighteen inches, and if it be poor, a good dressing of rotten manure or leaf mould should be mixed with it. Pears on quince stocks are much better adapted for this mode of culture than those on pear stocks. If the latter be planted, the border, six feet wide, should have a thick layer of concrete at bottom, to prevent the roots striking downwards; or it would be good practice to place, eighteen inches deep under each tree, a flat piece of stone, three feet in diameter – this would force the roots to take a horizontal direction, and facilitate the operation of root- pruning.

For fine specimens of wall pear trees grafted onthe quince, I may refer to those on the west wall ol the Royal Horticultural Society's Gardens at Chis- wick. These are now about forty years old, and are pictures of health and fertility, thus at once settling the question respecting the early decay of pear trees grafted on the quince; for it has been often, very often, urged as an objection to the use of the quince stock, that pears grafted on it are, although prolific, but very short-lived. I have seen trees in France more than fifty years old, and those above referred to may be adduced to confute this error.

1 If the wall to which the trees are trained be twelve feet and upwards in height, the border should be eight and even ten feet in width. Wide and shallow fruit tree borders are much to be preferred to those that are deep and narrow.

PEAK TBEES TRAINED AS SINGLE VERTICAL CORDONS.

The French gardeners have a curious yet interesting mode of training pears on the quince stock, about which a book was published in France a few years since. The system, I have recently learnt from some French cultivators, is now largely practised in the South of France with the peach apricot. It is called training ' en fuseau' or distaff training, and is the most simple of all modes. A young tree, one year old from the bud, is planted, and every *side shoot* as soon as it has made four or five leaves, is pinched to three. This is the first pinching, early in June. These pinched shoots all put forth young shoots, which must be pinched to one leaf; and so on with all the young shoots, during the summer, and the like practice every season. This will of course lead to spurs of some length being produced ; these spurs should in the second or third year be reduced to half their length, 'so as to keep the fruit-buds close to thestem. This reducing of the spurs should be done in winter, and is required in all cordon training. When the leading shoot has grown twelve inches, its top should be pinched off, and as soon as two or three break out at this point, all should be pinched in but one for the leader. A very compact distaff-like tree is thus formed.

For small gardens., where the cultivator wishes for a large collection of pears in a small place, this (which is, in fact, the cordon system applied to single Stemmed trees) is to be recommended.

Fig. 19 is a single cordon apple tree from a specimen growing here (single cordon pear trees require the same culture), and will, perhaps, give the reader a correct idea of the adaptability of these compact trees for small gardens ; they may be planted two feet apart.

DIAGONAL SINGLE CORDONS.

Fig. 12, the diagonal single cordon, is a pear tree grafted on the quince, planted and trained against a wall or fence at an angle of 45, the trees eighteen to twenty inches apart. They should be pinched in, and managed exactly as recommended for the vertical cordons. The gentle slope given to them seems to promote fruitfulness. Diagonal cordons of pears, plums, cherries, apples, and apricots, may be cultivated with success when trained against walls with south-west and all other aspects, except north or north-east. i

There is perhaps no wall-fruit tree so likely to be largely benefited by single diagonal training as the apricot. Every gardener knows the wretched disappointment often felt in summer by large and apparently healthy branches of their apricot trees dying off suddenly, and leaving them without aiiy remedy – for the gap made cannot be filled, owing to the rigidity of the remaining branches. There is, therefore, no remedy for this failure of apricot trees when trained to walls in the usual manner, but there is a sure method of avoiding it – simple enough : it is by

Fig. 12.

planting single diagonal cordon trees, which may be maiden trees with a single stem, or trees in a bearing state from the nursery. In planting, if the tree is slender, it is usual to keep the stem of the stock as nearly upright as possible ; but as the graft is often too stiff to bend readily, the tree may be planted slopingly.

Single diagonal apricot trees require a south or south-west aspect, and should be planted eighteen to twenty inches apart, and every shoot pinched in during the summer, as directed for cordon pear trees (p. 48), and the same directions as to reducing and thinning out the fruit-spurs in winter are necessary. The leading shoot need not, as a general rule, be shortened till it reaches the top of the wall, as the shoot of an apricot tree is generally so robust and full of buds. A single diagonal apricot tree, sloped to an angle of 45 or so, will, when it reaches the top of a wall 10 feet in height, be a cordon 15 feet in length. A wall 20 feet long will thus give space enough for ten or twelve trees, which in the course of two or three years will bear enormous quantities of fruit. One most important advantage, I repeat, is held out by this mode of culture, – no unseemly gaps need be seen, owing to the death of branches, as in the present mode, for whenever a tree dies – a very uncommon event, – it may be at once replaced. The expense of ten trees instead of one may be urged by the planter, costing 15s. instead of *7s. 6d.* for one well-trained tree. I have only to remark, that when the system is fully carried out the demand will be met by a much cheaper supply, and it must be recollected that it gives a ten-fold advantage over the old method of training.

Above all, it does away with the tiresome annual necessity of ' laying in' shoots, and pruning and nailing in winter. The single diagonal tree merely requires three or four stout shreds or, what would be a decided improvement, the same number of bands of india-rubber, to fasten the stem to the wall.

Peaches and nectarines trained as diagonal cordons against walls with a S. 'or S. W. aspect are worthy of a trial, but only in the warmer parts of England: here they are planted against the back wall of a lean-to house with an unshaded roof, and promise well.

The system of single diagonal training is so simple, all the pruning is pinching – that one feels assured of its being widely spread among amateur gardeners, who seem likely to lead the sound gardening taste of England. It must, however, be recollected that, although such trees trained against a wire fence are charming, they require protection from spring frosts, our great enemy.

The making of these wire fences for diagonal cordons is very simple. There are first straining posts of oak five inches by two and a half, these are placed firmly in the ground twenty feet apart, between these at six feet apart are the perforated flat slight iron bars used to support wire fences ; the wire may be stout iron wire the thickness of whipcord, which should be painted with coal-tar and lime, or if galvanized no painting will be required. The lowest wire is eighteen inches from the surface of the soil, and the other wires are one foot apart as high as required, but six, seven, or eight feet will be found high enough. Fig. 4, in Appendix, in which a wire is omitted by accident, will give an idea of diagonal cordon training on a wire fence. My trees are planted from fifteen to eighteen inches apart, and are models of beauty, far surpassing espalier training, and giving more fruit in the same space. For boundaryfences in the kitchen garden I know of nothing more desirable or more economical than a diagonal cordon fence, covered with trees carefully pinched and full of fruit.

One very happy idea occurs as I am writing this article: – The facility of protecting the trees while they are in bloom; for this purpose, where straw is cheap, straw mats could be employed or light frames covered with ' frigi domo ' or canvas, they should be made a little in excess of the height of the fence so as to overlap the trees ; thus, if the fence is five feet in height the covers should be six feet wide, and placed on each side of the trees to form a narrow span roof with the trees under it. The covers should be fastened to the top wire of the fence, and be removed daily, or at most, in bad weather with cold storms, once in two or three days. It must be recollected that cherries, plums, pears and apples may be cultivated after this fashion ; and judging by my cordon fences, now some three hundred yards in a very perfect state, they may be held as one of tho most popular methods of growing fruit.

PEAR-TREE HEDGE.

A few years since, when visiting a friend at Fontenay-aux-Roses, near Paris, I was much struck with a hedge formed of pear trees on the quince stock. He smiled when he told me his method of cultivation and pruning, the latter being simply clipping his hedge in July with the garden shears,1 and thinning out the spurs in winter, when they become crowded. A few days since (July, 1862), my friend paid me a visit, and I enquired, with some interest, about his pear-tree hedge. He assured me that it was perfectly healthy, and generally gave him large crops of fruit. The sorts proper to form a hedge are Louise Bonne of Jersey, Beurre d'Amanlis, Beurre Hardy, Conseiller de la Cour, Beurre d'Aremberg, Beurre Superfin, Doyenne du Cornice, Duchess d'Angouleme, and Vicar of Wiiikfield. These are all free growers on the quince stock, and if planted in a favourable soil and climate would soon form a fruitful hedge. They should be planted about thirty inches apart, and in masses, *i. e.,* planting, say, ten of each sort together. A hedge may be formed, varying more in its aspect by planting one or two trees of each sort in succession, – this is a mere matter of taste. A pear-tree hedge when in full bloom must have an agreeable look, and when full of fruit be very profitable.

1 An English cultivator would employ pruning scissors to shorten the h'Kts, and thus make his hedge look as if cared for.

PYRAMIDS ON THE PEAR STOCK.

There are some dry, warm, shallow soils, more particularly those resting on chalk or gravel, which are unfavourable to the pear on the quince stock ; it is difficult to make them flourish unless great care is taken in mulching the surface, and giving them abundance of water and liquid manure in summer. In such soils pyramids on the pear stock may be cultivated with but little trouble.

To those who wish to train them as they should grow, one-year-old grafted plants may be selected, which may be managed as directed for young pyramids on the quince stock. If trees of maturegrowth are planted, they will require the treatment recommended for pyramids on the quince stock, as regards summer pinching. There is no occasion, however, to make a mound up to the junction of the graft with the stock, as the pear does not readily emit roots. *Annual* root-pruning is almost indispensable to pyramids on pear stocks, in *small* gardens, and it will much facilitate this operation if each tree be planted on a small mound, the roots are then so easily brought to the surface. This annual operation, which should be done in November, may be dispensed with in soils not rich, if the trees be lifted biennially in that month and replanted, merely pruning off the ends of any long roots. Annual surface manuring, as recommended for pyramids on the quince, is also necessary, if the trees be root-pruned or biennially removed.

Trees of the usual size and quality may be planted, and suffered to remain two years undisturbed, unless the soil be rich and they make vigorous shoots (say eighteen inches in length) the first season after planting, in which case operations may then commence the first season. Thus, supposing a tree to be planted in November or December, it may remain untouched two years from that period ; and then as early in autumn as possible a circumferential trench, twelve inches deep, should be dug, and every root cut with the knife and brought near to the surface, and the spade introduced under the trees, so as to completely intercept every perpendicular root.

The treddle spade used in this part of Hertfordshire is a very eligible instrument for this purpose, as the edge is steeled ajid very sharp. The following year, the third from planting, a trench may be again opened at fifteen inches from the stem, so as not to injure the fibrous roots of the preceding summer's growth, and the knife and spade again used to cut all the spreading and perpendicular roots that are getting out of bounds. The fourth year the same operation may be repeated at eighteen inches from the stem : and in all subsequent root-pruning this distance from the stem must be kept. This will leave enough undisturbed earth round each tree to sustain as much fruit as ought to grow, for the object is to obtain a small prolific tree. I find that in the course of years a perfect mass of fibrous roots is formed, which only requires the annual or biennial operation (the former if the tree be very vigorous) of a trench being dug, and the ball of earth heaved down to ascertain whether any large feeders are making their escape from it, and to cut them ofi'. But it must be borne in mind that this circular mass of soil will in a few years be exhausted ; to remedy which, I have had left round each tree eighteen inches from the stem, a slight depression of the soil, or, in other words, the trench has not been quite filled in. This circular furrow I have had filled, in December and January, with fresh liquid nightsoil, covering it with a coat of burnt earth two inches thick, which has had a most excellent effect. Any other liquid manure would undoubtedly have been equally efficacious, but my soil was poor, and I thought it required strong manure. As it did not come in contact with the roots, no injury resulted from using such a powerful raw manure. I. cannot impress too strongly upon the reader my conviction of the strong necessity of applying lime or chalk to soils deficient in this deposit, I believe that many so- called exhausted borders require only the addition of lime in some form or other to renovate decaying trees.

Gas lime after an exposure to the air, superphosphate, lime rubbish or chalk will all be found to act beneficially.

There is no absolute necessity for liquid manuring in the winter, as common dung may be laid round each tree in autumn, and suffered to be washed in by the rains in winter, and drawn in by the worms. In mentioning liquid manure, I give the result of my own practice. The great end to attain seems (to use an agricultural phrase) to be able to ' feed at home;' that is, to give the mass of spongioles enough nutriment in a small space. A tree will then make shoots from eight to ten inches long in one season (for such ought to be the maximum of growth), and at the same time be able to produce abundance of blossom-buds and fruit. On trees of many varieties the former will be in too great abundance : removing a portion in early spring, cutting them out with a sharp knife, so as to leave each fruit-spur about three inches apart, is excellent culture.

I have not yet mentioned the possibility of root- pruning fruit trees of twenty or thirty years' growth with advantage. Irregular amputation of the roots of too vigorous fruit trees is, I am aware, an old practice ; but the regular and annual or biennial pruning of them, so as to keep a tree full of youth and vigor in a stationary and prolific state, has not, that I am aware of, been recommended by any known author, although it may have been practised. In urging its applicability to trees of twenty or thirty years' growth, I must recommend caution: the circular trench should not be nearer the stem of a standard tree than three feet, or if it be a wall tree, four feet, and only two-thirds of the roots should be pruned the first season, leaving one-third to support the tree, so that it cannot be blown on one side by the wind, and these of course must be left where they will best give this support. The following season half the remaining roots may be cut, or, if the tree be inclined to vigor, all of them; but if it gives symptoms of being checked too much, they may, on the contrary, remain undisturbed for one, or even two seasons. If, as is often the case in pear trees, the roots are nearly all perpendicular, the tree must be supported with stakes for one or two years after complete root-pruning.

The following extract from a letter recently received from C. Roach Smith, Esq., the archaeologist, is interesting, as showing the prompt effects of root- pruning of trees: – 'I have only been an horticulturist for three years ; I took to two very beautiful *old* pear trees, which must have cost no end of nailing, cutting, and staking. On inquiry, I found that one (a Summer Bon Chretien) had never produced more than *one pear* annually ; the other upon a north wall had *never* given a single pear. I could get no aid from any one what to do with those trees, and no book then accessible helped me. I reflected on the natural habit of the pear tree, and coming to the conclusion thatthe cause of barrenness was exuberance of roots, I resolved to cut them. Before the leaves had fallen, a friend sent me ' The Retired Gardener,' an old book, translated from the French. In it I found an account of some experiments made in England which fortified me in the resolution I had taken. The first year the Summer Bon Chretien ' produced nine fruit. I pruned the roots more closely, and this year (1859), in spite of the uugenial spring, I saved fifty-nine pears. The other tree yielded thirty-six, but of so vile a quality that I have re-grafted the tree. A large plum treated in the same way, produced the season after being root-pruned, 2000 fruit.'

It will not, perhaps, be out of place here to enumerate a few of the advantages of systematic root- pruning and removing or lifting of pear, apple, and plum trees, and of growing them as pyramidal trees and bushes.

1st. Their eligibility for small gardens, even the smallest.

2ndly. The facility of thinning the blossom-buds and in some varieties, such as Gansel's Bergamot, and other shy-bearing sorts, of setting the blossoms, and of thinning and gathering the fruit.

Srdly. Their making the gardener independent pf the natural soil of his garden, as a few barrowfuls of rich mould and annual manure on the surface will support a tree for many, very many, years, thus placing bad soils nearly on a level with those the most favorable.

1 This is one of our oldest varieties, and remarkable for being a very shy bearer.

4thly. The capability of removing trees of fifteen, or twenty years' growth with as much facility as furniture. To tenants this will, indeed, be a boon, for perhaps one of the greatest annoyances a tenant is subject to is that of being obliged to leave behind him trees that he has nurtured with the utmost care.

I feel that in judicious root-pruning and annual manuring on the surface, so as to keep our fruit trees full of short well-ripened fruitful shoots, we are all inexperienced. I am reminded of a wall in a neighbouring garden, covered with peach and nectarine trees in the finest possible health.

For more than twenty years a healthy peach tree was never seen in this garden, as the subsoil is a cold white clay, full of chalk stones. This happy change has been brought about by biennially pruning the roots of the trees early in autumn, as soon as the fruit is gathered; in some cases lifting the trees and supplying their roots with a dressing of leaf-mould, sand, and rotten manure, equal parts. Powdered charcoal, or the ashes of burnt turf and rotten manure, also make an excellent root-dressing for cold heavy soils ; but if the soil be dry and poor and unfavorable to the peach and nectarine, loam and rotten manure is the best dressing for the roots, and also for the surface.

PLANTING AND AFTER MANAGEMENT.

Pyramidal pear trees of from three to five years old on the quince stock, root-pruned, and full of blossom- buds, may now be purchased. Trees of this description should, if possible, be planted before Christmas; but if the soil be very tenacious, the holes may be opened in the autumn, and the trees planted in February ; the soil will be mellowed and benefited by the frosts of winter.1

Pear trees grafted on the quince stock offer a curious anomaly, for if they are removed quite late in spring – say towards the end of March, when their blossom-buds are just on the point of bursting – they will bear a fine, and often an abundant crop of fruit. This is sometimes owing to the blossoms being retarded, and thus escaping the spring frosts ; but it has so often occurred here when no frosts have visited us that I notice it – in fact, no trees bear late removal so well as pears on quince stocks.

a – *if.*

Fig. 13. – a, junction of the graft with the stock. 6, the point up to which the stock should be covered.

In planting pear trees on the quince stock, it is quite necessary that the stock should be covered upto its junction with the graft. This joining of the graft to the stock is generally very evident, even' to the most ignorant in gardening matters; it usually assumes the form as given in Fig. 13, *a.*

1 The roots of pear trees on the quince stock, and, indeed, of all root-pruned trees are very fibrous. In planting, it is good practice to give each tree two shovelfuls of fine earth or mould rather dry – to place it on the roots and shake the tree, so that the mould is mixed with the mass of fibrous roots. Before the soil is all filled in, three or four gallons of water should be poured in, so as to wash the earth into every crevice. The roots should not be crammed into a small hole. A tree with its roots eighteen inches in diameter, will require a hole 2J feet in diameter, and so on in proportion.

If the soil be not excessively wet, the tree may be placed in a hole, say three feet in diameter and eighteen inches deep, in the usual way, so that the upper roots are slightly above the level of the surface s the tree will always settle down two or three inches the first season after planting. Some of the light compost recommended in page 23 should be filled in, and the tree well shaken, so that it is thoroughly mingled with its roots. The compost must then be trodden down; and so far the planting is finished. The earth should then be placed round the stem, and formed into a mound, which should cover the stock *up* to, *but not above,* the junction of the graft with the stock, in order to encourage it to emit roots into the surface soil, and to keep it (the stock) from becoming hard and ' barkbound.'

As the mound will subside by the heavy rains of winter, presuming that the trees have been planted in autumn, fresh compost of the same nature must be added in spring, and every succeeding autumn. A quarter of a peck of soot, strewed on the surface in a circle three feet in diameter round each tree in March, is an excellent stimulant. The great object in the culture of the pear on the quince stock is to encourage the growth of its very fibrous roots at the surface, so that they may feel the full influence of the sun and air. The slight mounds recommended may be made ornamental if required by placing pieces of rock orflint on them, which will also prevent the birds scratching at them for worms ; but the stones selected must not be very large and heavy – they should be about the size and weight of a brick. In light friable soils, the mounds may be from three to four inches above the surface of the surrounding soil; in heavy retentive wet soils, from six to eight inches will not be found too high.

In soils of a light dry nature the pear on the quince requires careful culture, I therefore recommend the surface round the tree to be covered during June, July and August, with short litter, or manure, and to give the trees once a week, in dry weather a drenching with guano water (about one pound to ten gallons), which must be well stirred before it is used. Each tree should have ten gallons poured gradually into the soil; by this method the finest fruit may be produced; and as it is very probable that ere many years elapse, we shall have exhibitions of pears, this will be the mode to procure fine specimens to show for prizes. I must also here repeat that lime rubbish or chalk should be applied to soils deficient in calcareous deposit; I think that all fruit trees would be benefited by a biennial dressing of superphosphate; gas lime after an exposure of a month or two may be advantageously mixed with the surface dressing of manure.

Our oldest gardening authors have said, that ' pears engrafted on the quince stock give their fairest fruit;'and they are correct. It has been asserted that the fruit is liable to be gritty and deficient in flavour. I can only say, that from my trees, growing on a cold clayey soil, I have tasted fruit of Marie Louise, Louise Bonne of Jersey, and others, all that could be wished for in size and flavour.

1 A clerical amateur has informed me that this mulching or placing half-rotten manure one or two inches deep on the surface in a circle from two to three feet in diameter, and one and a-half inches deep, according to the size of the tree, will prevent pears cracking.

In the course of my experience and since the above recommendation to plant on mounds was written, I have found it good practice in *very dry* soils to plant pear trees on the quince stock with the junction of the graft just level with the surface, so as . not to require mounds round their stems. The first season they should have some manure on the surface, laid in a circle round the stem ; and the second year a shallow basin, two feet in diameter and four inches deep should be dug round the stem, and filled with some manure about half-rotten. This basin thus filled will keep moist even in the most dry and hot weather, and will become full of fibrous roots. This is also an excellent method of renovating pear trees that have exhausted themselves by bearing too abundantly or that appear unhealthy by their leaves turning yellow. In such cases, when the trees are of advanced growth, a basin of the same depth, but three or more feet in diameter, should be formed and filled with manure ; in all cases for this purpose this should be but slightly decomposed.

BUSH PEAR TREES FOR A MARKET GARDEN.

There are many sunny favourable spots which the amateur gardener may turn to profit accompanied with pleasure, simply by planting bush pear trees graftedon the quince stock. The plantation should be a sort of nursery, and for this purpose they should be planted in rows, four feet row from row, and four apart in the rows; a piece of ground planted after this method -will contain 2,722 trees per imperial acre.

By pinching every shoot to three leaves all the summer, the trees form compact fruitful bushes ; this constant summer pinching has a remarkable effect in moderating the vigor of fruit trees. They will commence to bear the second year after planting, and if each tree give but ten or twelve fruit, one acre will produce a large quantity. They may be suffered to remain at the above distance unroot-pruned, and un- removed for seven, eight or ten years ; and then, as they will nearly or quite touch each other, every alternate tree should be removed, and another plantation formed. The removal of the trees should be done carefully, so that those left will stand eight feet apart and in quincunx order, thus, . This may be done as follows : – Presuming the first row to consist of ten trees, begin at the first row by removing the 1st, 3rd, 5th, 7th, and 9th trees; in the second row, remove the 2nd, 4th, 6th, 8th, and 10th ; in the third row, again 1st, 3rd, 5th, 7th, and 9th trees, and so on with all, and through all the rows however long; at this distance they may remain for fourteen, eighteen or twenty years. At the end of one of these periods every alternate row of trees must be removed, leaving the permanent trees eight feet apart; the periods of removal must, to a certain extent, depend upon the nature of the soil; if this be of high fertility the removal of the trees must be commenced at the earlierperiod. It may sound strangely to the routine gardener to advise the removal of fruit trees when twenty years old; but I say this advisedly, for the trees in a plantation of Louise Bonne pears on the quince stock, planted here twenty years since in rows five feet apart, were recently removed and have succeeded well, commencing to bear fine crops the second season after being transplanted. When pyramidal or bush trees from ten to twenty years old are removed, their branches should all be shortened to at least one half their length. Although these trees were originally planted only five feet apart in the rows, and have grown well, they did not – and those left unremoved do not – touch each other; this is of course owing to their young shoots having been pinched in every summer for so many seasons.

From closely observing these trees for many years, and even to-day (July 20th, 1864), upon measuring the shoots of the unremoved trees, and finding they do not meet by at least fifteen inches, I have based the idea conveyed by the heading of these paragraphs p. 65. It may be asked, why not plant pyramids, which are handsome and productive ? Experience furnishes me with a reply ; when my 2000 pyramids of Louise Bonne pears commenced to bear their large crops of fruit, I found so many displaced by the wind that supporting them with stakes became expensive and troublesome ; I, therefore, recommend all those who wish to make their pear-tree plantations profitable as well as pleasurable, to plant bush trees.1 Insheltered gardens the amateur may, without hesita tion, continue to plant pyramids, for no description of fruit tree can be more interesting, but when profit is to be attached to cultivation, and fruit trees cultivated by the acre, the bush form should be adhered to. The varieties best adapted to this mode of culture are, first and best, Louise Bonne of Jersey, Beurre de l'As- sumption, Fondante d'Automne, Beurre d'Aremberg, Beurre Hardy, Beurre

Superfin, Williams' Bon Chretien, Souvenir du Congres, Beurre "Bachelier, Winter Nelis, Bergamotte d'Bsperen, Beurre Diel, Benrre d'Amanlis, Catillac, and Conseiller de la Cour; the plantation should be commenced with rows six feet apart, and the trees six feet apart in the rows. Iii the summer the weeds must be kept under by hoeing, which will keep the surface loose and promote the health of the trees, and, if the soil be poor, a dressing of manure in winter once in two or three years, leaving it to dissolve by the weather and the worms, for no digging in it must be allowed ; about five bushels of any kind of manure spread on twenty- five square yards will be the proper quantity, this will keep the surface tender so that it can be hoed to nourish the surface roots of the trees.

1 These may be with advantage a sort of hybrid bush tree, partaking a little of the pyramid, and allowed to grow to a height of four or five feet.

I ought here to mention, that amateur planters who think of planting bush pear trees in less quantities than I have alluded to, *i. e.,* by the dozen or score, may plant them 4 feet apart, row from row, and the same distance apart in the rows, as mentioned p. 65, but for permanent plantations 6 feet is to be preferred, cropping with light crops between the rows.

GATHERING THE FRUIT.

The fruit of pears, more particularly those on quince stocks, should not be suffered to ripen on the tree ; the summer and autumn varieties should be gathered before they are quite ripe, and left to ripen in the fruit room.1 The late pears should be gathered before the leaves take their autumnal tints ; if suffered to remain too long on the trees they frequently never ripen, but continue hard till they rot. In most seasons, from the beginning to the end of October is a good time; but much depends on soil and climate. The following passage from that very excellent work, Downing's ' Fruit Trees of America,' is appropriate to this subject:

' The pear is a peculiar fruit in one respect, which should always be kept in mind, viz., *that most varieties are much finer in flavor if picked from the tree, and ripened in the house,* than if allowed to become fully matured on the tree. There are a few exceptions to this rule, but they are very few. And, on the other hand, we know a great many varieties, which are only second or third-rate when ripened on the tree, but possess the highest and richest flavour if gathered at the proper time, and allowed to mature in the house. This proper season is easily known, first by the ripening of a few full-grown, but worm- eaten specimens, which fall soonest from the tree ; and secondly by the change of color, and the readiness of the stalk to part from its branch on gently raising the fruit. The fruit should then be gatheredor . so much of the crop as appears sufficiently matured, and spread out on shelves in the fruit room, or upon the floor of the garret. Here it will gradually assume its full colour and become deliconsly melting and luscious. Many sorts which if suffered to ripen in the sun and open air are rather dry, when ripened within doors are most abundantly melting and juicy. They will also last for a considerably longer period, if ripened in this way, maturing gradually as wanted for use, and being thus beyond the risk of loss or injury by violent storms or high winds.

1 Pears that ripen in September and October should not be "athered all at one time, but at intervals of a week or so, making, three gatherings ; their season is thus much prolonged.

' Winter dessert pears should be allowed to hang on the tree as long as possible, till the nights become frosty.1 They should then be wrapped separately in paper, packed in *Jeegs, barrels, or small boxes,* and placed in a cool dry room, free from frost. Some varieties, as the Beurre d'Aremberg, will ripen finely with no other care than placing them in barrels in the cellar, like apples. But most kinds of the finer winter dessert pears should be brought into a warm apartment for a couple of weeks before their usual season of maturity. They should be kept covered, to prevent shrivelling. Many sorts that are comparatively tough if ripened in a cold apartment, become very melting, buttery, and juicy, when allowed to mature in a room kept at a temperature of 60 or 70 deg.'

The following is from Mr. Glass's ' Gardening Book,' as given in the *Gardener's Chronicle:* −

1 I feel compelled to differ from Mr. D. in tins respect; for in the autumn of 1855,1 suffered many pears to hang on the trees till the end of October, and they never ripened. I believe the first week in October to be the best period to gather winter pears in.

HOW TO STORE WETTER PEARS IN SMALL QUANTITIES.

' Get some *unglazed* jars, − garden pots will do ; make them perfectly clean, if they have ever been used. The best way is to half burn ofr bake them over again.

' Gather your pears very carefully, so as not to rub off the bloom or break the stalk. On no account knock them about so as to bruise them. Put them on a dry sweet shelf, to sweat. When this sweating is over, rub them dry with a soft cloth, as tenderly as if you were dry rubbing a baby. ,

' As soon as they are quite dry, put them, one over the other, into the jars or garden pots, without any sort of packing; close up the mouth of the jar *loosely,* or of the garden-pot, by whelming the pan or placing a piece of slate over it, and stow them away in a darkish closet where they cannot get the frost.

' Open the jars now and then, to see how they are getting on.

' Do not put more than one sort in the same jar, if you can help it. Mind, − the warmer they are kept, the faster they will ripen.'

KEEPING PEARS IN A GREENHOUSE.

I have but very recently found that pears may be kept in a greenhouse, in great perfection, all the winter.

The greenhouse in which my experiment has been tried is a lean-to house with a S. W. aspect, twelve feet wide, with a path in the centre, a bench in frontof common slates laid on wooden bars, and a stage at back fall of camellias. My pears have been laid on the front bench, the glass over them shaded till the end of November, the house ventilated, and the camellias watered just as if the pears were not there. In severe frosts the temperature was kept just above freezing. The autumn pears under this treatment ripened slowly, and were of excellent flavour. The late pears kept till April; but then, owing to the power of the sun, the air of the house became too warm and dry, and they shrivelled. I should therefore recommend winter pears to be kept in the greenhouse in covered pots or jars (I now use large clean flower-pots with wooden covers), placing them in them early in December.

Mr. Tillery, of the Wellbeck Gardens, keeps his choice pears and apples in boxes of bran with great success. The bran before it is used should be thoroughly dried and sifted, so as to take from it all the small particles of meal. "With this treatment pears and apples may be placed in it as soon as they are gathered. The boxes should be quite shallow, so as to admit of only one layer of fruit, which should be covered with the bran and no lids placed on the boxes. The bran is apt to become musty.

After all, I think there is no better material for preserving pears plump and sound than dry cocoa- nut fibre : this never turns musty, never ferments, but seems to remain under all circumstances perfectly innocuous. To preserve pears in this substance they should be placed in two layers in shallow open boxes, first placing a layer of fibre two inches thick in a

box, then a layer of pears, which should he covered with the fibre, say one inch thick; on this place another layer of pears, and cover them to a depth of two inches. The fruit should be occasionally examined to ascertain when they are ripe.

PYRAMIDAL APPLE TREES ON THE PAEADISE APPLE STOCK.

Apples as pyramids on the Paradise stock are objects of great beauty and utility. This stock, like the quince, is remarkable for its tendency to emit numerous fibrous roots near the surface, and for contracting the growth of the graft, causing it to become fruitful at a very early stage. On the Continent, there are two varieties of the apple under this denomination, viz., the Doucin and the Pomme de Para- dis ; these are called Paradise stocks in England, but on the Continent the first and last are used for distinct purposes – the first for pyramids, the latter for dwarf bushes.

The Doucin stock is, I am inclined to think, the same as that called 'Dutch Creeper,' or 'Dutch Paradise,' by Miller, in his Dictionary, folio edition of 1759. It puts forth abundance of fibrous roots near the surface of the soil, and is not inclined to root deeply into it like the crab. Apples grafted on this stock are more vigorous than when grafted on the French Paradise stock, and less so than those 011 the crab ; it is, therefore, well adapted for garden trees, for they are easily lifted, their roots thus kept to the surface, and the tree consequently kept free from canker. There is another surface-rooting apple also well adapted for stocks, the Burr Knot. This, like the Doucin, will strike root, if stont cuttings, two or three years old, are planted two-thirds of their length in a moist soil; it is a large, handsome, and very good culinary apple. At Ware Park in Hertfordshire, this is called Byde's Walking-stick Apple, owing to Mr. Byde, the former proprietor of the place, often planting branches with his own hand, which soon formed nice bearing trees.

Among apples raised from seed, some will occasionally be found with this surface-rooting nature; and this is, I suspect, the reason why the Doucin stock, under the name of the Paradise, in the English nurseries, differs from the stock used as Doucins in France; there are also several varieties cultivated there, some of which are unfitted to our climate.

About forty years since, I raised a large number of apples from the pips of the Golden Pippin, Golden Reinette, Ribston Pippin, and other esteemed sorts. These, in course of time, all bore fruit, but as not one was found superior to its parent, I did not cultivate them. Why I mention this is, that among my seedlings were several that put out roots near the surface, and the cuttings of which struck root. It is only within these few years that I have had my attention drawn to two of these, one of which has very broad leaves, and a most healthy and vigorous habit; the other, a habit equally vigorous, but with a great tendency to form fruit spurs. The former I have named the Broad-leaved Paradise, the latter, the Nonesuch Paradise ; – they are likely to form a revolution in apple culture, as the varieties of apples grafted on them formed such healthy and fruitful trees.

I have at this moment (Sept., 1870) a full collection of all the Paradise stocks known in Europe. There are three varieties of the French Paradise, all making very dwarf trees; then come three Dutch Paradise, all much alike, but slightly more vigorous than the French sorts ; next to them are two English Paradise, both of them from old English nurseries – they have much resemblance to the French Doucin stock, but are better, swelling with the graft. The Creeping Paradise is probably that mentioned by Miller, in the last century, since it is very remarkable for putting forth suckers from the roots, objectionable, but not common with the apple tribe. The Nonesuch Paradise stock raised from that very old apple the Nonesuch of Queen Bess's time, is quite *sui generis,* for it has downy leaves and a knotted stem, but is wonderfully fertile. The Broad-leaved Paradise, also raised from seed here, is much like the best varieties of the Doucin stocks, of which there are endless varieties; one of the best in my plantation is good and much like the last named. The Miniature and Pigmy Paradise, both raised from seed here, give that very dwarf habit which the French Paradise does. 1 have thus enumerated fourteen kinds of Paradise stocks, the three first and the two last remarkable for giving very dwarf trees; all the others are of the same nature as the Doucin, all giving dwarf healthy trees.

The Pommier de Paradis, or the French Paradise, seems identical with the ' dwarf apple of Armenia,' referred to in the 'Journal of the Horticultural Society,' Part 2, Vol. 3, page 115. It is *exceedingly* dwarf in its habit, and too tender for this climate, unless in very warm and rich soils. Out of 2,000 imported in 1845, more than half died the first season, and two-thirds of the remainder the following. They were planted in fine fertile loam, favourable to the growth of apples, and on which the Doncin, planted the same season, grew with the greatest vigour. The same result attended an importation in 1866. I have now potted some plants and owing, as I suppose, to the roots being warmed through the pots by exposure to, the sun, they seem inclined to make very nice little fruitful bushes, – in fact real miniature apple trees, bearing fruit when only nine inches in height. My trees are in eight-inch pots ; but to have healthy fertile trees, I should recommend them to be gradually shifted into fifteen-inch pots. The citizen may thus have his apple orchard on the leads of his house.

The English Paradise stock, much like the Doucin, and those before-mentioned as my seedlings, are most deserving of our attention as stocks for forming fruitful healthy pyramids and bushes, the culture of which is very simple. Grafted trees of one, two, or three years' growth, with straight leading stems, well furnished with buds and branches to the junction with the stock, should be planted. No manure should be placed to their roots, but some light friable mould should be shaken into them, the earth filled in, trodden down, and two or three shovelsful of halfrotten manure laid on the surface round each tree. This surface-dressing may be given with advantage every succeeding autumn. If the soil be very wet and retentive, it will be better to plant the trees in small mounds; and if symptoms of canker make their appearance, their roots should be examined annually in the autumn, as recommended in root- pruning of pears on the quince stock, introducing the spade directly under the roots, so as to prevent any entering deeply into the soil, and bringing all as nearly to the surface as possible, filling in the trench with light friable compost; or the tree may be lifted and replanted, which will be found more efficient. I firmly believe that canker may be entirely prevented by this annual attention to the roots.

If, therefore, the soil be unfavourable, and apt to induce a too vigorous growth in apple trees, followed by canker, the roots should be annually root-pruned, or the trees lifted – i. e., taken up and replanted. If, however, the trees make shoots of only moderate vigour, and are healthy and fruitful, their roots may remain undisturbed; and pinching their shoots in summer, as directed for pyramidal pears, page 9, and training them in a proper direction, is all that they will want. Pyramids on the Paradise stock may be planted four feet apart in confined gardens; five feet will give them abundance of room ; but if, owing to the soil being of an extra fertility, they are found to require more, the trees, if they have been root-pruned, may be removed almost without receiving a check, even if they are twenty years old. This is a greatcomfort to the amateur gardener, who amuses himself with improving his garden ; for how often does a favourite fruit tree, which cannot be removed, prevent some projected improvement!

Apples differ greatly in their habits of growth ; some are inclined to grow close and compact, like a cypress – these are the proper sorts for pyramids; others, horizontally and crooked – these should be grown as bushes ; others again are slender and thin in their growth, so that, to form a good pyramid of these slender-growing varieties, it is necessary to begin the first year with a young tree, and to pinch the leader as soon as it is six inches long. If by any neglect the lower part of the pyramid be not furnished with shoots, but have dormant buds, or buds with only two or three leaves attached, a notch must be cut, about half an inch in width, just *above* the bud from which a shoot is required. The notch must be cut through the outer and inner bark, and alburnum, or first layer of wood ; and if the shoot or stem be young – say from two to four inches in girth – it may be cut round half its circumference. If this be done in spring or summer, the following season a shoot will generally make its appearance ; sometimes even the first season, if the stem or branch be notched early in spring. This method of producing shoots from dormant buds may be applied with advantage to all kinds of fruit trees, except the peach and nectarine, which are not often inclined to break from a dormant bud.

Varieties of apples, inclined to be compact and close in their growth, form very handsome pyramids ; but they are apt to be unfruitful, as air enough, is not admitted to the interior of the tree. This may be easily amended, by bringing the lateral shoots down to a horizontal position for a year or two, and fastening the end of each shoot to a stake; an open pyramidal shape will thus be attained, which the tree will keep. Other varieties put forth their laterals horizontally, and some are even pendulous. The leading perpendicular shoot of varieties of this description should be supported by a stake, till the tree is of mature age. Iron rods, about the size of small curtain-rods, are the most eligible: these, if painted with coal-tar and lime, sifted and mixed with it to the consistence of very thick paint, put on boiling hot, will last a great many years.

Apple trees in confined gardens near large towns, are often infested with ' American blight,' *aphis lanigera:* this makes its appearance on the trees generally towards the middle of summer, like patches of cotton wool. There are many remedies given for this pest; the most efficacious I have yet found is soft soap dissolved in soft water, two pounds to the gallon, or the Gishurst Compound, sold by Price's Candle Company, one pound to the gallon, and applied with an old painter's brush. Many remedies such as train oil, spirits of tar, &c. are apt to injure the trees : it must be recollected that soft soap will turn the leaves brown – in fact, kill them ; but it need not be applied to them, as the *aphis* generally fixes itself on the branches.

Here let me impress upon the lover of his garden, living anywhere within the reach of smoke, thenecessity of using the syringe : its efficacy is not half appreciated hy garden amateurs. As soon as the leaves of his fruit trees are fully expanded, every morning and every evening, in dry weather, should the attentive gardener dash on the water with an unsparing hand – not with a plaything, but with the perforated common syringe, such as a practical gardener would use, capable of pouring a sharp stream on to the plant, and of dislodging all the dust or soot that may have accumulated in twelve hours. For apple and pear trees in pots, or in small city gardens, this syringing is absolutely necessary.

Pinching the shoots of pyramidal apple trees, and indeed, exactly the same method of managing the trees as given for pyramidal pears on the quince stock, may be followed with a certainty of success ; and the proprietor of a *very* small garden may thus raise apple trees which will be sure to give him much gratification. To have fine fruit the clusters should be thinned in June; and small trees should not be overburdened, for they are often inclined, like young pear trees on the quince stock, to bear too many fruit when in a very young state ; the constitution of the tree then receives a shock which it will take two or three seasons to recover. For varieties with large fruit, one on each fruit-bearing spur will be enough ; if a small sort, from two to three will be sufficient.

There are so many really good apples that it is difficult to make a selection : the following sorts will not disappoint the planter; but fifty varieties in addition, quite equal in quality, could be selected.

Twenty dessert apples, ripening from July to June, placed in the order of their ripening : –
1. White Joanneting
2. Early Red Margaret
3. Red Astrachan
4. Early Strawberry
5. Irish Peach
6. Summer Golden Pippin
7. Kerry Pippin
8. Margil

9. Ribston Pippin
10. Cox's Orange Pippin
11. Mannington's Pearmain
12. Golden Drop (Coe's)
13. Ashmead's Kernel
14. Nonpareil, Old
15. Reinette Van Mons
16. Syke House Russet
17. Keddleston Pippin
18. Golden Harvey
19. Melon
20. Stunner Pippin
Twenty kitchen apples, fit for use from July to June: –
1. Keswick Codlin
2. Lord Suffield
3. Hawthornden
4. Cellini
5. Cox's Pomona
6. Blenheim Pippin
7. Calville Blanche
8. New Hawthornden
9. Striped Beefing 10. Mere de Me"nage
11. Herefordshire Pearmain
12. Winter Pearmain
13. Bedfordshire Foundling
14. Small's Admirable
15. Dumelow's Seedling
16. Betty Geeson
17. Rymer
18. Baxter's Pearmain
19. Warner's King
20. Gooseberry Apple.

APPLES AS BUSHES ON THE PARADISE STOCK.

There are some varieties of apples that do not form, even with care, well-shaped pyramids ; such sorts may be cultivated as bushes when grafted on the Paradise stock, and are then excellently well adapted for (Small gardens. I have, indeed, reason to think that a great change may be brought about in suburban fruit culture by these bush trees. I have shown in pp. 20 and 21, how bush pears on quince stocks may be cultivated. Pears are, however, a luxury ; apples and plums are necessaries for the families of countless thousands living near London. Apple bushes, always very pretty and productive trees, may be planted fourfeet apart, row from row, and four feet apart in the rows. If two or three years old when planted, they will begin to bear even the first season after planting. They should be kept from the attacks of the green aphis in summer by dressing the young shoots with the quassia mixture, given in a note to p. 113, and from the woolly aphis by Gishurst Compound, men-

Fig. 14.

tioned in page 78. The principal feature in this culture is summer pinching, which must regularly be attended to, from early in June till the end of August; this is done by pinching or cutting off the end of every shoot as soon as it has made five or six leaves, leaving from three to four *full-sized* ones. Some varieties of the apple have their leaves very thicklyplaced on the shoots ; with them it is better not to count the leaves, but to leave the shoots from three and a-half to four inches in length. If the soil be rich, and the trees inclined to grow too vigorously, they may be removed biennially, as recommended for bush pears, by digging a circular trench one foot from the stem of the tree, and then introducing the spade under its roots, heaving it up so as to detach them all from the soil, and then filling in the earth dug from the trench and treading it gently on to the roots. The following sorts are well adapted for this bush culture, but the upright varieties recommended for pyramids form nice compact bushes.1

Brabant Bellefleur, kitchen .... April
Cornish Aromatic, dessert .... May
Early Harvest, dessert ..... August
Emperor Alexander, kitchen .... October
Gravenstein, kitchen or dessert . . . November
Cox's Orange Pippin, dessert .... October
Cox's Pomona, kitchen ..... December
Lord Burghley, dessert . . . . - . May
" Hawthornden, kitchen
Joanneting (White), dessert .... July
Melon Apple, dessert ..... February
Mere de Me"nage, kitchen .... December
Nonesuch, kitchen ...... October
Reinette du Canada, kitchen or dessert . . May
Ribston Pippin, dessert ..... December
Spring Ribston Pippin, dessert . . . May
Waltham Abbey Seedling, kitchen. . . December

There is no mode of apple culture more interesting than bush culture. On page 84 I annex a sketch of a plantation of Cox's Orange Pippin (Fig. 15), of one hundred trees ; they were planted in the spring of1862. They bore a fine crop in 1863 of most beauti- fal fruit, and in 1864 gave a crop almost too abundant.

1 These dwarf bushes are liable to be gnawed by rabbits and hares in exposed gardens. The best of all preventatives is to paint them with soot and milk, well mixed ; or make a fence with galvanized wire netting, round the garden in which they are planted. .

APPLES AS BUSHES FOR MARKET GARDENS.

In a well-ordered fruit garden every kind of fruit should have its department, and instead of seeing, as in Kent, a row of trees of all sorts, mixed in the most heterogeneous manner, no mixture of species should be allowed ; every kind should have its allotment, – apples on the Paradise stock, ditto on the crab stock, pears on the quince stock, the same on the pear stock. Morello cherries as pyramids on the Mahaleb stock – the best of all methods for their culture – and the various kinds of the Duke cherries on the same kind of stock. Heart and JBigarreau cherries on the common cherry stock, plums as bushes, pyramids, or half standards, should all be separated, and not planted higgledy-piggledy, as they have been and are now. The sound-headed market gardener will, when his mind is turned to improved fruit-tree culture, see all this and make his fruit-garden a pattern of order.

I have been led into these remarks on market garden fruit-tree culture by my own experience, and especially into a consideration of the great improvement that may be made in the culture of apples on the English Paradise stock. On referring to p. 82, the reader will find that I allude to my plantation of Cox's Orange Pippin apple trees on the Paradise stock (see Fig. 15) ; these trees in the season of 1864, the third of their growth in their present quarters, and the fourth of their age, gave an average of a quarter of a peck from each tree, so that we might have from 4840 trees, growing on one acre ofan acre of apple trees a very agreeable and eligible investment. The kinds likely to sell best in the markets, and which are most productive, are the following: – Cox's Orange Pippin, Ribston Pippin, Stunner Pippin, Scarlet Nonpareil, and Dutch Mignonne ; these are dessert apples. The following are valuable kitchen apples, and abundant bearers: – Hawthornden, New Hawthornden, Small's Admirable, Cox's Pomona, Keswick Codlin, Dumelow's Seedling, Lord Suffield, Norfolk Bearer, and Duchess of Oldenburgh. Such large varieties as Bedfordshire Foundling, Blenheim Orange, and Warner's King, should have more space, be planted four feet apart, and be thinned out by removal to eight feet apart as recommended for pear trees. The proper method of planting and managing these bush apple trees, is exactly that recommended for bush pear trees on quince stocks.

ground 302 bushels of fine apples worth 5s. per bushel, or 75Z. In 1866, the trees then averaging half-a-peck each, would double this sum, and make

It may be, by some, made a question of expense, for although the return must be large and profitable, the purchase of nearly 5000 apple trees would involve a large outlay. To this I reply – first, that stocks costing only a small sum per 1000 may be planted and grafted where the trees are to grow permanently; and, secondly, that a large demand which my method of planting would create, will also create a cheap supply. The preparation of an acre of ground should be as follows : – It should, previous to planting, be forked over to the depth of twenty inches (if very poor and exhausted, from thirty to forty tons of manure may be forked in), – not more, as trees such as I have recommended, viz., pears on the quince stock, and apples on the English Paradisestock do not root deeply – this ought to cost 12*l*. The annual expenses are, forking the surface in spring, *ll. 6s. 8d.,* and hoeing the ground, say four times during the summer, 1Z. 4s. I give the amounts paid here for such work. Then comes the summer pinching of the shoots by a light-fingered active youth, and this may, at a guess, be put down at 1Z., making the aggregate annual expenses, 3Z. 10s. *Sd.,* or, say, 4Z. per acre. The large return will amply afford this outlay, even adding, as we ought to do, the interest on capital, and rent.

It will be seen that what I propose is in reality a Nursery Orchard which may be made to furnish fruit and trees for a considerable number of years. To fully comprehend this we must suppose a rood of ground planted, as I have described, with 1210 bush apple trees. In the course of eight or ten years half of these, or 605, may be removed to a fresh plantation, in which they may be planted six feet apart; they will at once occupy half an acre of ground. At the end of sixteen or eighteen years every alternate row of trees in the first plantation – the rood – will require to be removed, which will give 302 trees to be planted, six feet apart, leaving 303 in the original rood. The 1210 trees will by this time occupy one acre of ground at six feet apart. With proper summer pruning or pinching they will not require any further change, but continue to grow and bear fruit as long as they are properly cultivated. The great advantage reaped by the planter is the constant productiveness of his trees ; from the second year after planting they will be always ' paying their way.'

The unprejudiced fruit cultivator will quickly find out the great advantage of my mode of apple and pear cultivation. Still, it may be thought too serious a business to attend to three or four thousand trees per acre, and only adapted to a very humble cultivator ; I ought therefore to state that those who wish to cultivate apples and pears for market purposes may with a sound prospect of success, if the soil and climate are favourable, plant apples on the English Paradise stock, and pears on the quince stock, either as pyramids or bushes, four and six feet apart, row from row, the former distance for dwarf prolific sorts, the latter for robust growers. This distance will admit of light crops of vegetables for two feet in the centre between each row for several years, and till the trees – which must be under summer pinching – cover the ground. I

In the usual old-fashioned mode, Standard apple trees are planted in orchards at 20 feet apart, or 108 trees to the acre ; if the soil be good and the trees properly planted, and the planter a healthy middle- aged man, he may hope at the end of his threescore and ten, to see his trees commencing to bear, and may die with the reflection that he has left a valuable orchard as a legacy to his children, but has not had much enjoyment of it during his life. Now, although, like most fathers, I have a strong wish to benefit my children, I hold the idea that one ought also to think of one's own gratification; and so I have planted and recommend the planting of such trees as will give me *some* satisfaction, yet leave a fertile inheritance to my children. H

A French pomologist, who paid me a visit in 1864, said, ' Ah ! now I find an Englishman planting for himself as well as for his children ;' and went on to say that he was struck by seeing in England so many Standard trees in market gardens, the planters of which could have derived but small benefit from them ; and an apparent ignorance of fruit gardening as a lucrative occupation. This he, in fact, imputed to our climate, which, Frenchman-like, he thought totally unfit for fruit culture in the open air, yet felt much surprised to see here the produce of a well cultivated English fruit garden, in a climate not nearly so favourable as the valley of the Thames.

I have only to add, that besides my plantation of Cox's Orange Pippin, 1 have another of upwards of 400 trees, which has now been in existence upwards of twenty years, so that I am not theorising but deducing facts from a sound basis. Since writing the above, my attention has been directed to a plantation here; its apple trees are grafted on the English Paradise, they are four feet apart and are safe for twenty or thirty years if pinched and pruned as directed.

Fig. 16.
APPLES AS SINGLE LATERAL CORDONS.

The French gardeners often train an apple tree ' en cordon horizontal,' as an edging to the borders in their kitchen gardens, after the following mode: A tree grafted on the Paradise or Doucin stock, with a single shoot, is planted in a sloping position, and the shoot trained along a wire, about ten or twelve inches from the surface. (Fig. 16.)

To carry out this methodof training, oakposts, about three inches in diameter and two feet in length, should be sharpened at one end and driven into the ground, so that they stand one foot above the surface; they may be from thirty to forty yards distant from each other.

From these a piece of galvanized or common iron wire – if the latter it should be painted – about the thickness of whipcord, should be strained and supported nine inches from the ground at intervals of six feet, by iron pins eighteen inches long, the size of a small curtain rod, or smaller, flattened at top and pierced with a hole, to allow the wire to pass through ; these should be stuck into the ground, so as to stand on a level with the straining posts. The trees should be planted six feet apart, and when the top of one tree reaches to another the young shoot may be grafted on to the base of the next, so as to form a continuous cordon. This is best done by merely taking off a slip of bark, two inches long, from the under part of the young shoot, and a corresponding piece of bark from the upper part of the stem of the tree to which it is to be united, so that they fit tolerably well. They should then be firmly bound with bast and a branch of moss – a handful – as firmly bound over the union ; the binding as well as the moss, may remain on till autumn. The trees do not grow so rapidly as common grafts, so that the ligatures will not cut into the bark.

4

SECTION 4

Every side shoot of these cordons should be rigorously pinched down to three leaves all the summer. It will of course occur to the reader, that the spurs

would soon make the tree a thick and clumsy cordon; to prevent this, every spur should be reduced in winter to about half its length, and some of the crowded blossom-buds removed with a very sharp

knife. The fruit, from being near the earth, and thus profiting largely by radiation, will be very fine. As these low cordons are very apt to be injured in winter by severe frost, if snow is suffered to lie under them, which by resisting radiation gives great intensity to frost just above its surface, it is necessary either to carefully remove the snow, to bank it up so as to completely cover the cordons, or to thatch them with a covering of evergreen branches, such as furze, or of firs, fern would also be a safe protection – better than all, are wooden ridges made of -inch boards, so as to cWertwo or three rows of trees. For pear trees there should be boards on one side and glass on the other, they would then do to protect the blossom in spring, and bring on the fruit if placed on bricks as directed for ground vineries.

The double, or two-branched lateral cordon, see Fig. 17, which is a great improvement on the French single cordon, requires the same training, pinching- in and management. This improved lateral cordon does not require a wire to support its branches, a kind of hook, something after a shepherd's crook may be used with advantage, thus : – the branch is introduced at *a* and is supported by the crook.

The quadruple lateral cordon is a tree well adapted for the edging of the borders of the kitchen garden ; it is merely the double cordon repeated, and we must suppose the two branches of the double cordon to be trained nine inches from the surface of the ground, and above them, at about nine inches distance, two

other branches in the same direction ; this will give the quadruple cordon (Fig. 18), or low espalier edging trees, occupying no more space than the single cordon, and giving double its produce. The stem of the short crook for single or double cordons should be 20 inches long, that of the longer one, for quadruple cordons, should be 28 inches long.

The great change in fruit culture that may be brought about by training these, double lateral cordons under glass ridges is obvious enough. The figure (19) will give some faint idea of the advantages of this new system of culture – they are endless ; for not only can peaches, nectarines, apricots, plums, apples and pears be rescued from spring frosts, but their fruit be ripened in great perfection. There is no doubt but that in some of our cold and cloudy places in the north of England and Scotland, where even the Ribston Pippin will not ripen, it may be brought to perfection under the glass fruit ridge.

The figure (19) gives but one tree trained to one wire ; two rows of wire may, however, be trained under one glass ridge, which should be three feet six inches wide at base, and the wires ten inches asunder. It is quite possible that this method of training to galvanized wires may, in some situations, be better adapted to vine'culture than allowing the vines to rest on slates or tiles.

I now, by permission, copy the description of my new glass fruit ridge from my article in the *Gardeners' Chronicle* for April 8, 1865, from which I have also derived the plate kindly lent to me :

' There are no cross bars, but merely a frame three feet wide at the base. On the top bar a, is a groovehalf an inch deep ; in the bottom bar *l,* is a groove a quarter of an inch deep ; *l* in the end bars *a* and *d,* are grooves half an inch deep. The pieces of glass, which should be cut so as to fit, are pushed into the upper groove, and let fall into the lower one ; when all are fitted in, the two end pieces are pushed inwards, so as to drive all of them into close contact. A little putty is required at the bottom to prevent water lodging, and some at each end to keep the pieces from moving laterally, *e, e,* are the straining posts of oak, four inches square ; *l,* the upright pieces of wire stuck in the ground, flattened and perforated at top to pass the wire through and support it: *g,* the wire.'

Such, then, is the description of the barless glass fruit ridge, which I think calculated to have a, greater effect on domestic gardening, and contribute more to the refinement and comfort of a very large class of people than all the crystal palaces ever invented. I feel that I ought to add how and where these nice things are to be bought.

Mr. James Rivett, builder, of Stratford, Essex, makes and sells them at 5s. *6d.* or 6s. each, unglazed. Those who would wish to have a large number, and who live at a long distance from London, should have a few from Mr. Rivett as samples; they could then be imitated by any good labourer.

For ventilation and other particulars I refer my readers to the description of the ground vinery, p. 141 ; and for the method of placing the wires, to p. 93.

I must caution those who wish to grow fruit underglass fruit ridges, in small confined gardens, to be careful as to ventilation. A single row of bricks, with apertures of four inches, will not be enough ;

1 An improvement on this is to have a rebate at bottom instead of a groove ; the glass is more easily fitted in.

there should be two rows of bricks, one over the other, and consequently two rows of apertures. Peach, nectarine, and apricot trees should be planted twenty-one feet apart; but they grow rapidly, and would probably require occasional removing.

It will thus be seen that to commence glass fruit ridge culture, three seven-feet lengths should be prepared, and in the centre of the twenty-one feet occupied by the ridge, two peach or nectarine trees may be planted. They will soon form lateral cordons of great fertility, will require pinching weekly, and give constant employment to the amateur. I must not omit to state the great advantage this mode of fruit culture gives as to protection from spring frosts when the trees are in bloom, or when the fruit is young. Espaliers, pyramids and wall trees are difficult to protect, but mats two or three thick can be piled on the ridge with great facility, and loose straw or hay, the best protectors possible from frost, can be strewed over them thickly.

I had, in the season of 1870, the pleasure of seeing all my anticipations fully realised; the cordon pear-trees have produced fruit large and with the fine clear rinds we see on those grown in the warm parts of France – perfectly beautiful and of fine flavour. The cordon peach-trees have produced fruit large and of the finest flavour. Strawberries planted between the trees temporarily till they fully occupy the room under the ridge, ripened a fortnight earlier than those in the open air, and were of excellent quality. I have, therefore, no hesitation in recommending this mode of fruit culture to all amateurs who have gardens without walls or orchard houses.

VERTICAL CORDON APPLE TREES.

In pp. 48 and 49 will be found the method of training vertical cordon pear trees. This may be applied to apples on the English Paradise stock with great success, and very charming fruitful trees they make. They should not be allowed to grow above eight feet in height, to which they will reach in the course of four or five years. I annex a figure of one of these trees, three years old, and full of fruit (Fig. 20a).

APPLES AS WALL TREES.

We have been so accustomed to think of, and treat the apple tree as hardy and perfectly adapted to our insular climate, that the culture of superior varieties as wall trees, has been neglected, except in the extreme north of our island, where the climate is not very favourable even to the culture of the Ribston Pippin as an orchard tree.

The varieties most worthy of cultivation against walls in England, even in our most favoured counties with regard to climate, are mostly of American origin, the continental varieties, with but very few exceptions, not being remarkable for goodness of quality.

The best methods of cultivation are : –

1. To have the trees trained as espaliers to low walls as directed for pear trees (p. 41), the trees to be under summer pinching as given p. 32. 2. To plant five-branched

upright cordons in the spaces so often found between wall-trees in old gardens. 3. To plant single vertical cordons (see Fig. 20a) against walls between established wall-trees ; these should have all the shoots and spurs cut off closely on one side of the trees, so that the stem may be easily fastened to the wall with a band Ojt even two or three strong shreds. Single vertical cordon apple treesgrafted on the English. Paradise stock and planted against walls 10 to 12 feet high, the trees well managed by summer-pinching, become amazingly prolific, and bear the finest of fruit. 4. To train at the foot of a wall the single lateral cordons (Fig. 16), or the double lateral cordons (Fig. 17) ; if the space next the wall and under the trees be paved with tiles or slates, the size and quality of the fruit will be improved. I ought here to mention that double or two-branched lateral cordon pear trees are to be preferred ; they may be grown at the foot of walls, but not more than 9 inches from them ; the tile-paving is with pears quite necessary, as is also protection in spring from frosts. This is most effectually done by lean-to barless lights in foot of the glass span ridge (Fig. 19) divided into two ; these most convenient lean-to lights should be 2 feet 4 inches wide, including the top and bottom bars, and seven feet long; two hooks should be fixed to the top bar, and two eyes in the wall, so that the lights are made safe from the effects of the wind. The lower bar should rest on bricks (they should be two deep), as with ground vineries. These lean-to lights will be found a most useful invention ; they form so fine a climate against brick

Fig. 20o.

walls, that I see no reason why low 4-inch brick walls, should not be built by market-gardeners, and lean-to lights of increased size employed for early crops ; the climate they give is perfect, so efficient is the low admission of cool air in between the bricks, and the exit of the heated air at the top between the upper bar and the wall, an interstice of about two inches.

The varieties of apples most worthy of wall culture are of American origin, viz., the Newtown Pippin, Washington, Bar's Apple, Melon, Northern Spy, Fall Pippin, Lady's Sweeting, and some others.

The French Apple, Calville Branche, is also of high excellence, cultivated as a wall or orchard house tree, and in cool climates, our fine English apples the Golden Pippin and Ribston Pippin are quite worthy of a place against a wall with a southern aspect.

PYRAMIDAL APPLES ON THE CRAB STOCK.

In soils light and poor, the apple on the Paradise stock is, unless carefully manured on the surface, apt to become stunted and unhealthy. In such soils and also in those of a very tenacious nature, pyramids on the crab stock may be planted with great advantage. They are also well adapted for large gardens where large quantities are required, as the trees may be made to form handsome pyramids, from twelve to fifteen feet in height.

Carefully watch the trees, for there is one thing most essential to their full success as pyramids, – they must either be lifted or taken up biennially early inNovember, and replanted in the manner recommended for bush pear trees, or root-pruned biennially, operating upon the trees alternately, as mentioned in note to p. 14; or the following system may be adopted : neither remove nor root-prune any tree that continues to grow with moderation, does not canker, and bears well; but any tree that makes shoots from eighteen inches to three feet in length, remove once in two, three, or four years till its vigorous habit is reduced.

As these crab stock trees grow freely, summer pinching or shortening the young shoots with a penknife as recommended in p. 81, must be attended to, and then, in the most unfavourable apple-tree soils, healthy and most prolific pyramids may be formed. Any of the varieties recommended in p. 82 will succeed well as pyramids on the crab stock.

If managed as I have directed, fine trees may be formed, not only of the robust-growing kinds, but even of the old Nonpareil, Golden Pippin, Golden Reinette, Hawthornden, Ribston Pippin, and several others, all more or less inclined to canker. I have a row of Nonpareils and Ribston Pippins planted in the coldest and most unfavourable soil I could find, yet, owing to their being biennially removed, they are entirely free from canker.

The vigorous growth of standard apples, when planted in orchards in the usual way, is well known, and also their tendency to canker after a few years of luxuriant growth. Pyramids on the crab, without occasional removal, or root-pruning, would, in like manner, grow most freely ; and even if subjected to summer pinching, would soon become a mass of entangled, barren, cankered shoots.

PYRAMIDAL PLUM TREES.

. The plum, if planted in a rich garden soil, rapidly forms a pyramid of large growth, – it, in fact, can scarcely be managed by summer pinching. It becomes crowded with young shoots and leaves, and the shortening of its strong horizontal branches at the end of summer is apt to bring on the gum: it is a tree, however, with most manageable roots, for they are always near the surface. I must, therefore, again recommend summer pinching to three leaves, as directed for pears, p. 9, annual or biennial root- pruning, and surface dressing, in preference to any other mode of culture. The root-pruning of the plum is performed as follows : – Open a circular trench eighteen inches deep round the tree, eighteen inches from its stem, and cut off *every root and fibre* with a sharp knife. When the roots are so pruned, introduce a spade under one side of the tree, and heave it over, so as not to leave a single tap-root; fill in your mould, give a top dressing of manure, and it is finished. The diameter of your circular trench must be slowly increased as years roll on; for you must, each year, prune to within one-and-a-half or two inches of the stumps of the former year. Your circular mass of fibrous roots will thus slowly increase, your tree will make short and well-ripened shoots, and bear abundantly. From very recent experience, I have found that removing trees annually, if the soil be rich – biennially, and adding some rich compost, if it be poor, – *without root-pruning,* will keep plum trees in a healthy and fertile state. For further particulars on this head, see pp. 13 and 14.

Pyramidal plum trees are most beautiful trees both when in flower and fruit. Their rich purple and golden crop has an admirable effect on a well-managed pyramid. No stock has yet been found to cramp the energies of the plum tree. I have, however, tried experiments on the sloe, which, as it never forms a tree of any bulk, effects this object to a certain extent. My trees on the sloe are some years old, and are dwarf and prolific. The first year after grafting they made vigorous growth ; but this is a very common occurrence with stocks that ultimately make very prolific trees ; it is so with the pear on the quince, the apple on the Paradise, and the cherry on the Mahaleb. The greengage seems to grow more freely on the sloe than any other sort. I have three fine vigorous bushes, now about ten years old, growing in the white marly clay, with chalk-stones, peculiar to some parts of Essex and Hertfordshire. The sloe seems to delight in this soil, so inimical to most kinds of fruit trees. My greengage plums are almost vigorous in their growth; and what appears strange is, that the stock seems to keep pace with the graft – there is scarcely any swelling at the junction. The roots of these trees have not been touched, and they appear to have gone deeply into the solid white clay. The plum on the sloe is easily arrested in its growth by root pruning. I have some trees, four years old, not more than eighteen inches high, and yet covered with blossom I

buds.1 These have been only once root-pruned, and are forming themselves into nice compact prolific bushes. As no peculiar culture, or disease, requires to be noticed, I have only to give a selection of sorts calculated for pyramids. These are also well adapted for walls with W., N. W., E., or S. E. aspects.

HARDY DESSERT PLUMS ADAPTED FOR PYRAMIDS. *In season from July to the end of October. Placed in the order of their ripening.*
Early Favorite
Early Greengage
De Montfort
Oullin's Golden Gage
Greengage
Jefferson
Kirk's
Transparent Gage
Purple Gage
Guthrie's Late Green
Reine Claude de Bavay
Bryanstone Gage

HARDY KITCHEN PLUMS ADAPTED FOR PYRAMIDS.
In season from July to the end of October. Placed in the order of their ripening.
Early Prolific ! Mitchel. son's
Sultan Diamond
Belgian Pnrple
Pond's Seedling
Prince Engelbert
Victoria or Alderton
Imperial de Milan
Autumn Compote
Late Black Orleans
Belle de Septembre

PLUM TREES AS BUSHES.

There is, perhaps, no fruit tree so easily kept within bounds as the plum. In rich soils they bear annual removal with but a slight check ; but in most soils biennial removal will keep them in a perfectly fruitful state under bush culture. This is absolutely necessary; and if the soil be poor, some thoroughly rotted manure (about half a bushel to each tree) may be mixed with the soil in replanting. As with pear trees, the best season for lifting or removing them isthe end of October or beginning of November. Plum bushes have the advantage of being easily protected by a square of light cheap calico, tiffany, or any light material thrown over them while in blossom, and a crop of fruit thus insured. All the varieties recommended for pyramids may be cultivated as bushes, and for suburban gardens, they should be subjected to exactly the same treatment as recommended for apple bushes, p. 81.

1 Since this was written, I have found plums grafted on the plum stock so easily dwarfed by annual or biennial removal, that unless in hard clayey soils, found to be unfavourable to the plum, there is no occasion to employ the sloe stock, unless as an experiment.

PLUM TREES AS COEDONS.

The plum forms a most prolific lateral double cordon and gives very fine fruit, when pruned and trained after the fashion of pear trees. Owing, however, to the fruit often receiving injury from heavy rains, it is almost indispensable to have a space under each tree paved with tiles, and a work of necessity to protect the trees from spring frosts, for they (the trees) come into blossom so early, owing to their receiving the reflected heat from the soil in early spring, that seldom or never does the young fruit survive the month of April. One of the best modes of protection are those ridges of glass and boards described p. 93, for if placed on bricks, they may remain over the trees till the commencement of the first week in June, here a period of rejoicing, for not till then are we safe from the fruitgrower's scourge – a severe spring frost. There is a method of cultivating a few kinds of plums as vertical cordons practised here which is likely to be popular, it is simply selecting the proper sorts (first catching i2 your hare), and then planting them in ground not too rich, – say a calcareous sandy loam, and then pinching in, during the summer, all the young shoots as directed p. 48, and trusting to this, to restrain the growth of the trees, without either root-pruning or removal.

The varieties adapted to this mode of culture are as yet but few, viz.: – Oullins' Golden Gage, Reine Claude de Bavay, Belgian Purple, Sultan, Cluster Damson and Prince Engelbert, of the latter kind upwards of 1000 trees are planted here for fruit bearers, they are now five years old, and are becoming compact, fertile cypress-like trees. In the course of time, there will doubtless be many kinds of plums adapted to this mode of culture, for here we have 2000 seedling plums all raised from choice varieties, and likely to give kinds as well adapted to our climate, as is the Early Rivers or Early Prolific, the hardiest plum known, but yet only the first removed by seed, from one of the most tender French' varieties, Precoce de Tours plum.

These vertical cordon plums should be planted from 4 to 5 feet apart, row from row,, and the same distance tree from tree : the former distance will allow of 2700 trees per acre, the latter 1700, and as far as I can see, many years will elapse before they will require thinning, and they will bear many bushels of fruit per acre.

MARKET GARDEN PLUM TREES.

Plums, like pears, open up a rich field to the amateur market gardener, for it is found that theyare so easily made into articles of exportation, by jam and bottling, that the demand is limitless.

The same method of culture as given for pyramidal pears on the quince stock (p. 21) is at once the most simple and beneficial.

The trees may be planted 6 feet apart, row from row, and 6 feet apart in the rows ; for a few years the centre of the spaces between the rows may be cropped with light crops. I grow strawberries, but onions or other light crops of vegetables may be grown. As soon as the treee have made sufficient growth to shade the ground, which may be in five or six years, more or less, the ground should have a dressing of manure, and be left undug ; the hoe only, to kill the weeds, should be employed. The following kinds will be found the best for this mode of culture : – Early Prolific, Sultan, Prince Engelbert, Belgian Purple, Reine Claude de Bavay, Cluster Damson and Belle de Septembre. The second sort of the last- named is so pyramidal in its growth that it may be planted 3 feet apart in the rows, and every alternate tree removed at the end of seven to ten years.

The Autumn Compote and Victoria, two very hardy useful plums, may be planted 6 feet apart as directed, but their stems will require a stake to each to support them for some three to four years, or till they become stout enough to stand without support.

Cherries As Bushes And Pyramids On The Mahaleb Stock (cerasus Mahaleb).

This stock has been long known in our shrubberies as the ' Perfumed cherry;' its wood when burnedemits a most agreeable perfume. In France it *is* called 'Bois de Ste. Lucie,' and it has been used there for dwarf cherries for very many years ; – why it has not been employed by English nurserymen, I cannot tell. My attention was called to it in France some twenty or more years ago, since which I have used it extensively, annually increasing my culture. Its great recommendation is, that cherries grafted on it will flourish in soils unfavourable to them on the common cherry stock, such as strong white clay or soils with a chalky subsoil. Although the trees grow most vigorously the first two or three seasons, yet after that period, and especially if root-pruned, they form dwarf prolific bushes, so as easily to be covered with a net, or, what is better, with muslin or tiffany, which will protect the blossoms from frost in spring, and the fruit more effectually from birds and wasps in summer ; thus giving us, what is certainly most rare, cherries fully ripe, and prolonging their season till September. These dwarf bushes may be planted from five to six feet apart, and their branches pruned so that seven, or nine, or more, come out from the centre of the plant, lik, e a well-managed gooseberry bush. These branches will, in May or June, put forth, as in the horizontal shoots of pyramidal pears, several shoots at their extremities, all of which must be pinched off to three leaves, leaving the leading shoots untouched till the middle or end of August, when they must be shortened, and the pruning for the year is finished.

The Morello and Duke cherries – the most eligible for this bush culture – may have their leading shooJsshortened to eight leaves. If, however, the space be confined in which they are planted, this length may be reduced, for by biennial root-pruning, the trees may be kept exceedingly dwarf. The aim is to form the tree into a round bush, not too much crowded with shoots. Towards the end of September, or, in fact, as soon as the autumnal rains have sufficiently penetrated the soil, a trench may be dug round the tree, exactly the same as recommended for root- pruning of pears, the spade introduced under the tree to cut all perpendicular roots, and all the spreading roots shortened with the knife, and brought near to the surface, previously filling in the trench with some light friable soil for them to rest on, and spreading them regularly round the tree, as near to the surface as possible; then covering them with the soil that was taken out of the trench. No dung or manure of any kind is required, as this stock seems to flourish in the poorest soils. Some short litter, or half decayed leaves will, however, be of much benefit placed on the surface round the stem.

I have thus far given their culture for small gardens ; but those who have more space may dispense with the root-pruning, and allow their cherry trees to make large bushes, which may be planted eight feet apart and pinched regularly in the summer, and managed as directed for pear trees (p. 10). The leading shoot from each branch in such cases must be left longer, and shortened to twelve or more buds.

1 This early autumnal root-pruning will be found very advantageous, the flow of sap is checked, so that the shoots are well ripened, and the roots soon emit fresh fibres to feed the tree the following season.

The most charming of all pyramids are the varieties of the Duke and Morello cherries on the Mahaleb ; these by summer pinching, as practised for pyramidal pears, become in two or three years the most delightful fruit trees ever seen, for in spring they are perfect nosegays of flowers, and in summer clusters of fruit – if spared by spring frosts.

The common Morello cherry on the Mahaleb stock, cultivated as a pyramid, forms one of the most prolific of trees ; but as birds carry off the fruit when only half ripe, each pyramid should have a bag of tiffany placed over it, and tied round the stem of the tree at bottom. Any garden, however small, may grow enough of this useful sort by planting a few pyramids, lifting and replanting, or root-pruning them biennially and pinching in *every slwoi* to three leaves (as soon as it has made five) all the summer. The Kentish cherry, also a most useful culinary sort, may be cultivated as a pyramid with great success. A French variety grown near Paris, in large quantities, and known as the ' Cerise Aigre Hative,' which may be Englished by calling it the Early Sour Cherry, is a useful kind for the kitchen. In going from Paris, a year or two ago, to Versailles by the ' Rive Droite ' Railway, I was much struck by seeing in the market gardens between Suresnes and Puteaux, on the left, large plots of dwarf trees, about the size of large gooseberry bushes, and some very low trees, all covered (as they appeared to me from the railway carriage) with bright red flowers. I learned, on enquiry, that these were cherry bushes – literally masses of fruit, of the above variety. I

From a Photograph, August, 1862. *Kg.* 21.

find, however, that it is not equal to the Kentish in flavour or size in England.

I need scarcely add, that the culture of all the Duke tribe of cherries by closely pinched-in pyramids, biennially removed, or biennially rookpruned, is most satisfactory. It is, perhaps, more easily performed than root-pruning, and the trees soon form perfect pictures. As far as my experience has gone, cherries on the Mahaleb are much more fruitful when ' oft removed ;' the most eligible mode is to remove only half the trees in one season, and the remainder the following season. I have seen nothing in fruit- tree culture more interesting than handsome compact pyramids of such sorts of cherries as the May Duke, Duchesse de Palluau, Empress Eugenie, and Archduke. One feels surprised to find that as yet but few lovers of gardening know of the existence of such trees. It will much facilitate the operation on their roots if the trees be planted on small mounds.

In forming plantations of pyramidal and dwarf cherries on the Mahaleb stock, it is necessary to arrange them with a little care. The two groups, those of the habit of the Morello tribe, and those of the compact habit of the May Duke, should be planted in separate rows. Bigarreau and Heart cherries are too shortlived in many kinds of soil, when grafted on this stock – unless double-grafted on the Morello cherry – to be recommended. The following arrangement will assist the planter: –

SECTION I. – The May Duke Tribe.

Archduke
May Duke
Royal Duke
Jeffrey's Duke
Belle de Choisy
Nouvelle Royale
Empress Euge'nie
Duches? e de Palluau

SECTION II. – Thr Morrllo Tribe.

Carnation (Coe's) late
Kentish
Late Duke
Morello
Eeine Hortense
Planchoury

Cherries grafted on the Cerasus Mahaleb are eminently adapted for espaliers, or for walls, as they occupy less space, arid are very fertile. They may be planted twelve feet apart, whereas espaliers on the cherry stock require to be eighteen or twenty feet apart. For potting, for forcing, cherries on this stock are highly eligible, as they grow slowly and bear abundantly.1

CHERRIES AS SINGLE VERTICAL CORDONS.

The varieties best adapted for this very interesting mode of culture are those of the Duke tribe, such as the May Duke, Archduke, Empress Eugenie, Royal Duke, Nouvelle Royale, Duchesse de Palluau, and some others. Young pyramidal trees three feet apart should be planted in rows, and their side shoots pruned into within two inches of their stems. They require the same summer pinching as that recommended for vertical cordon pears, p. 48, and should not be allowed to exceed eight or ten feet in height. Nothing can be more charming than these cordon cherry trees. I have at this moment trees five years old, of the Duke tribe, with their bright ripe fruit hanging close to the stem, and shining through the net that protects them from the birds.2 The best ofall protection, both from birds and wasps, is, however, Haythorn's netting, or coarse muslin, formed into a narrow bottomless bag, which should be let down gently over the tree, so as to leave the leading shoot out, and tied at the bottom and top ; Duke cherries may thus be preserved till August. I may mention here, that with these cherry cordon trees, root- pruning or removal is seldom required, their vital force is so reduced by the continuous pinching of the young shoots; but if a rich soil gives too much vigour, it may be practised. There are a few kind of plums, of upright growth, which may also be cultivated as vertical cordons.

1 Cherry trees are often infested in summer with the black aphis. The best remedy is a mixture made by boiling: four ounces of quassia chips, in a gallon of soft water, ten minutes, and dissolving in it, as it cools, four ounces of soft soap. It should be stirred, and the trees syringed with it twice or thrice. The day following they should be syringed with pure water. "2 Written in July, 1870.

The Bigarreau and Heart, or Guigne cherries, are too vigorous for this mode of culture when grafted or budded, as they generally are, on the common cherry stock. The new mode of culture by double grafting, i. e., by grafting them on young trees of the common Morello cherry that have been grafted on the Mahaleb, will make them most prolific cordons. (See p. 128.)

I must add a piece of very necessary advice : all vertical cordon trees, whether pears, apples, cherries, or plums, should be supported by a slight iron rod, about the size of a goose-quill, which should be painted; this should stand six to seven feet above the surface, and be inserted ten to twelve inches in the ground, and the tree attached loosely to it by two or three bands of sheet lead or some soft metal.

EIGARREAU AND HEART CHERRIES AS PYRAMIDS ON THE COMMON CHERRY STOCK.

Among the mysteries of vegetable physiology

there, is nothing, perhaps, more interesting than the facts discovered by the fruit-cultivator. Many kinds of pears grow with great luxuriance when grafted or budded on the quince stock, while other kinds, cultivated in the same soil, and budded or grafted with equal care, will grow feebly, and die in the course of a year or two.

The Noblesse and Royal George peaches form fine healthy trees when budded on the Mussell Plum stock.

The Grosse Mignonne and the French Galande die in a year or two, if budded on it. The Moor Park apricot grows readily and freely on the above-named stock. The peach apricot, its French congener, will not; why ? The Bigarreau and the Heart cherries (or, as the French call them, Guignes) do not succeed so well on the Cerasus Mahaleb as they do on the common cherry ; they grow most rapidly for two or three years, and then are apt to become diseased.

The stock raised from the small black and red wild cherries is the proper one for this race, except they are double-grafted.

Pyramidal cherry treeo may be bought ready-made or formed, by purchasing young trees, one year old, from the bud, and training them up in the same way as directed for pyramidal pears (pp. 5 and 6), with this variation, – pears, as is well known, may be grown as pyramids successfully, with or without root- pruning or biennial removal; but cherries on common cherry stocks will grow so rapidly, in spite of summer pinching, that biennial removal is a work of necessity. In the course of a few years pyramidal cherry trees thus treated become pictures of beauty. In Francethey generally fail, and become full of dead stumps and gum, owing to their trusting entirely to pruning their trees severely in summer and winter, without attending to their roots; the trees thus being full of vigour make strong shoots, only to be pinched and cut off. We must ' manage these things better in England.

The mode of operation in removing pyramidal cherries is the same as that recommended for pears and apples, &c. It will be found, however, that more labour is required, for in two years the cherry on the common stock, like the apple on the crab, makes a vigorous attempt to lay hold of its parent earth. The second year the tree may be lifted by digging a trench round its stem, one foot from it and 116 inches deep. The fourth year this trench must be 18 inches from the stem and 20 inches deep; the sixth year it should be 2 feet from the stem and 2 feet deep. This distance and depth need not be departed from if the trees are required to be only fair- sized pyramids ; the straggling roots beyond this circumference should be biennially pruned off with the knife. The tree managed thus, will soon be in a mature fruitful state, and its roots a mass of fibres, so that when removed it will, like the rhododendron, receive only a healthy check.

Pyramidal Bigarreau and Heart cherries, cultivated after the method above given, may be planted in small grass orchards, with pyramidal pears on pear stocks, pyramidal apples on crab stocks, and pyramidal plums. A charming orchard in miniature may thus be formed. Cattle and sheep must, ofcourse be excluded, and a wire fence, enclosing a space from three to four feet in diameter, should be round each tree. This space must be kept free from grass and weeds.

The following varieties form handsome pyramidal trees, and bear fruit of the finest quality : –

Bigarreau Jaboulny
Bohemian Bigarreau
Large Black Bigarreau
Early Black Bigarreau
Late Purple Guigne
garreau
Black Tartarian
Down ton
Elton
Florence
Governor Wood
Werder's Earlv Black
Bigarreau Napole"on

I have thus far given the results of my experience in the culture of pyramidal trees. The method is not by any means new, for visitors to the Continent, for these last fifty years, must have often observed the numerous pyramids of France and Belgium. The system of annual and biennial root-pruning I must, however, claim as original, for I feel assured that in our moist climate – too moist for many varieties of fruit – such a check is required to keep pyramids that are under summer pinching in a healthy, fruitful state. The defect in the pyramidal trees of the Continental gardeners, is their tendency to an enormous production of leaves and shoots, brought on by severe annual pruning of their shoots. The climate is probably too dry for root-pruning ; yet

I cannot help thinking that if it were followed by manuring thickly on the surface and occasional watering, it would make their trees prodigiously fruitful,

At the risk of repetition, and writing from my own experience, I must say that no gardening operationcan be more agreeable than paying daily attention to a plantation of pyramids. From the end of May to the end of July – those beautiful months of our short summer – there are always shoots to watch, to pinch, to direct, fruit to thin, and a host of pleasant operations, so winning to one who loves his garden and every tree and plant in it.

To conclude, I may mention that the small Alberge apricot, raised from the stone, and producing small high-flavoured fruit, and also the Breda apricot make very beautiful pyramids if lifted or planted biennially. In the southern counties of England, in a favourable season, they will ripen their fruit, and produce good crops. The large Portugal quince is also very prolific as a pyramidal tree. Some trees only two years old have borne fine fruit here. This is the finest of all the quinces, and in the south of Europe it grows to an enormous size. The Medlar will also form a handsome and productive pyramid, and, ' last, but not least' in the estimation of the lover of soft fruits, the currant. A near neighbour – an ingenious gardener – attaches much value, and with reason, to his pyramidal currant trees; for his table is supplied abundantly with their fruit till late in autumn. The leading shoots of his trees are fastened to iron rods; they form nice pyramids of about five feet in height; and by the clever contrivance of slipping a bag made of tiffany over every tree as soon as the fruit is ripe, fastening it securely to the bottom, wasps, and birds, and flies, and all the ills that beset ripe currants are excluded. With all these, summer pinching and root-pruning, or occa-

sional removal (except the currant, which does mot require the latter operation), as directed for pears, are indispensable; they soon form very handsome pyramids, and make a pleasing variety in the fruit garden.

FILBERTS AND NUTS AS STANDARDS.

Filberts, as commonly cultivated, except in the Kentish gardens, form straggling bushes, and are some years before they commence to bear. To correct this, I some ten or more years since had them grafted by inarching on stems of the hazel-nut raised from Spanish nuts, as they were vigorous growers and formed stout stems. I have found these grafted trees answer admirably, and come quickly into bear ing, forming nice garden trees.

As soon as the nut trees designed for stocks, have made stout stems about four feet high, they should be grafted by inarching at that height with choice kind of nuts, such as the red and white filberts and the Cosford nut – an excellent nut. The purple-leaved filbert, generally planted as an ornamental shrub, may also be grafted ; it gives nuts equal to the common filbert, and forms a nice ornamental standard.

Standard nuts require but little culture; they soon form round heads, and bear profusely. Care must be taken to destroy all suckers from the stem and root.

The only pruning required is in winter to thin out the crowded shoots, and shorten to half their length those that are inclined to be vigorous – that is, those that are more than nine inches in length. The short spray-like shoots must not be shortened, as they are the fruit-givers.

Standard nuts planted in rich garden soils soon make trees too large for small gardens. If, therefore, they are found to grow too vigorously, they should be lifted and replanted biennially in November.

I have mentioned seedling nuts as good for stocks ; but I have lately employed a valuable sort, introduced from Germany as Corylus arborescens; this makes a beautiful clear stem.

The Algiers nut, Corylus algeriensis, seems also to be well adapted for a stock for standards, as it makes shoots from six to seven feet in one season. It also produces very fine and handsome fruit.

FIGS AS HALF, STANDARDS OR BUSHES.

There is, perhaps, no fruit tree that disappoints the amateur fruit grower so much as the fig. If planted in the open borders of the garden, it soon grows into an enormous fruitless bush or tree, and if placed against a wall, unless a very large space can be given to it, but little fruit must be expected.

It may, however, be made eligible for small gardens, where the climate is sufficiently warm to ripen its fruit, such as the gardens near London, and those in the eastern and southern counties. Fruitfulness and moderate growth are brought on by the following method. Trees should be procured of the An- gelique, Brown Turkey, White Marseilles, and Early Violet Figs – these are the only kinds that bear freely, and ripen their fruit well – such trees should be low or half standards, or dwarfs with a clear stem(not bushes branching from the ground). The former should have a stem three feet high, and the latter from one foot to eighteen inches; in each case the tree should have a nice rounded head.

Trees thus selected should be planted in a sunny situation, and require only the following simple mode of treatment. They, we will assume, were planted in March or April. They will make a tolerably vigorous growth, and must be pruned by pinching off the top of every shoot as soon as it has made six leaves, leaving five. The stem must be kept quite clear from young shoots. By the autumn nice round-headed trees will be formed, and about the end of October they should be taken up (their leaves cut off if they have not fallen) and placed in a cellar – no matter if dark, but a light dry cellar would be preferable – some earth should be placed over their roots, and there they may remain till the first week in May, when they should be planted out, and the same routine of culture followed. They will bear one good crop of fruit in a season and ripen it in September. This annual removal brings on great sturdiness of growth in the tree, and the roots become so fibrous as to hold a large quantity of earth, which should not be shaken from them when they go into their annual winter abode. In the year 1857 I saw fine trees thus treated in the garden of the Duke of Altenburg, in Central Germany, their stems were as stout as a man's leg, and their heads full of fruit; and this season, 1865, my fig trees, taken up in October, and placed in the orchard-house during the winter – theirroots in the soil – gave me a crop of very rich, well- ripened fruit. The sorts were the Brown Ischia, Brown Turkey and Brunswick.

THE BIENNIAL REMOVAL OF FRUIT TREES WITHOUT ROOT-PRUNING.

For some few years past I have felt a growing conviction that peach trees trained agaist walls in the usual manner, without careful root-cultivation, cannot, in our climate, be kept in a state at all healthy or fertile for a series of years. A wall covered with healthy peach, or nectarine trees of a good ripe age is rarely to be seen ; failing crops and blighted trees are the rule, healthy and fertile trees the exception. The following mode of treating peaches, nectarines, apricots, and plums on the removal system I have found simple and efficacious: – Supposing a trained tree, of the usual size, to hare been planted in a border well prepared – *i. e.,* stirred to a depth of twenty inches; it may be trained to the wall as usual, and suffered to grow two seasons. Towards the end of October, or, indeed, any time in November in the second season, it should be carefully taken up, with all its roots intact. If there be two or three stragglers – *i. e.,* roots of two or three feet in length – for roots are remarkably eccentric, and often, without any apparent cause, run away in search of something they take a fancy to – cut off one foot or so, so as to make the roots of the tree more snug. Then make the hole from which you took your tree a little deeper, and fit to receive its roots

without bending or twisting. If the soil be heavy, leaf-mould, or rotten manure, and loam, -equal parts; if it be light, two-thirds tenacious loam, and one-third rotten manure, should form the compost; two inches deep of this, placed at the bottom of the hole, will be enough for the roots of the tree to rest on ; and mind they are carefully arranged, so as to diverge regularly: then add enough of the compost to cover all the roots, and fill in with the common soil, so as not to cover the surface roots more than two inches deep. The surface should be trodden down very firmly, so as to be like a path, and then have a dressing two inches deep of old tan or decayed litter ; add lime rubbish or chalk if a non-calcareous soil.

A tree that has been planted two years will require one barrowful of the above compost; at the end of four years two barrowfuls; when six years have passed from three to four barrowfuls; and from four to six barrowfuls will be enough for a tree from twelve to twenty years old – in short, for a full-grown tree. A portion of the earth from the border must be removed when a large quantity of compost is added, to make room for it, so as not to have an unsightly mound. In the course of two or three removals the roots of the tree will become a mass of fibres, and the trees so docile as to be lifted without difficulty.

I have this day (Dec. 12,1852) removed two plum trees that have been planted six years and removed twice. Their roots are a mass of fibres, without one straggling root; they have been replanted witha barrowful of light compost to each tree,1 and if I may judge by the enormous quantity of blossom- buds, they will bear a plentiful crop next season: They will receive no unhealthy check, for abundance of earth adheres to the mass of fibrous roots. Now, as peaches, nectarines and apricots, being budded on plum stocks, are all on plum roots, they will give exactly the same results from the same mode of culture, neither the *size nor flavour* of the fruit will be affected, and the trees will always bear abundantly, and be healthy and flourishing.

The plethoric habit of the Moor Park and Peach apricots, which so often leads to disease and death will be effectually cured by this simple mode of culture, and peaches and nectarines will make short annual shoots, which will be always well ripened, so that they will be constantly full of healthy blossom-buds. Some mulch, or old tan, two inches in depth, placed on the surface of the soil – which should always be trodden down firmly – as far as the roots spread during the spring and summer will be of much service.

All trees that are inclined to make very fibrous roots, such as plums, pears or quince stocks, and apples on Paradise stocks, may be lifted – *i. e.,* removed biennially or occasionally, if their growth is not too vigorous, as above described – with equal or greater facility than root-pruning them.   The effect is the same ; they make short well-ripened shoots, and bear abundantly.   Apples on Paradise stocks, cultivated as dwarf bushes or as pyramids, if lifted *every year,* and a shovelful or two of compost given to them, form delightful little trees.1 The most delicate sorts of apples, such as Golden Pippins and Nonpareils, may thus be cultivated in the most unfavourable soils ; and Roses, more particularly Bourbon Roses on short stems, and Hybrid Perpetnals, removed annually in the autumn, giving to each tree a shovelful of rich compost, and not pruning their shoots till April, will bloom delightfully all the autumn, never dropping their leaves towards the end of summer, and not becoming, as is too often the case, blighted and blossomless.

1 The soil is rich, and one barrowful I thought quite enough.   The quantity of compost must be regulated by the wants of the soil, for in rich soils, where peaches and nectarines are apt to grow too freely, no compost need be added, but the tree merely lifted and replaced.   A peach, nectarine, or apricot tree, under the removal system, that makes annual shoots more than fifteen inches in length, is too luxuriant, and will require no compost to its roots when replanted.

To conclude, I will, as a guide to the amateur owning a small garden, give the following summary : – If the soil be very rich, so as to induce the trees planted in it to make a growth of eighteen inches in one season, they may be removed *amwally* till this vigorous growth ceases. If the trees make an annual growth Only of ten to twelve inches, the trees may be removed *biennially or occasionally,* and I may add that in soils *in which trees grow slowly,* root-pruning is more advaatageous than removal, as less check is given to vegetation.

1 In moist retentive soils the fruit-spurs of small trees become covered with moss; some powerful lime sprinkled over them will destroy it; this is best done in foggy weather in winter.

DOUBLE GEAFTING OF FRUIT TREES.

I have not been able to find this mode of culture, likely to be so beneficial to fruit gardens in England, alluded to by the many authors of works on fruit trees ; it may be ' as old as the hills,' and have no claim to originality, but few so-called new ideas have. I can only therefore state how it originated here some fifteen or twenty years since. I am not aware that it has been practised by the clever fruit tree cultivators of France and Belgium ; if so, it has been recently copied from English practice,' but I never remember having seen it carried out. It may be described in a very few words. A double grafted pear tree is formed by selecting a variety that grows very freely when budded or grafted on the quince, and re-grafting it, *i. e.,* grafting the graft with a kind that refuses to unite kindly with the quince stock.

Its history, briefly told, is as follows : – I observed when budding and grafting pears on the quince stock, that some varieties did not grow freely on that stock, when budded or grafted; particularly the Gansel's Bergamot and the Autumn Bergamot, the

Seckle, the Marie Louise, Knight's Monarch, and some others. Now, as the first and last mentioned are notorious for their shy bearing qualities, while the trees are young, even when root-pruned, or frequently removed, I felt anxious to see them, flourishing on the quince stock, which invariably makes pear trees fertile. I found that but few grafts of these sorts out of scores would survive on the quince, and when they did unite they were very? short-lived; this induced me to look narrowly into the habits of pear trees on the quince stock, and I found that the Beurre d'Amanlis formed a most perfect union with the stock, and seemed most enduring, for I had seen trees in France at least fifty years old. I therefore fixed upon this sort for my experiment, and had thrifty trees two years old from the bud, grafted with Gansel's Bergamot; the grafts flourished, and became so prolific, that when three or four years old, they each bore from three to four dozen of fruit – a most unusual thing with that fine variety. This settled the question as to the fertility given by double grafting; which since this experiment, has become here an extensive branch of culture. There are other kinds of pears which, from uniting with, and growing freely on, the quince stock, serve well for double grafting, such as Prince Albert, Bezi d'Antenay, and Conseiller de la Cour. Prince Albert is a sort well adapted for the Monarch, Marie Louise, Prince of Wales (Huyshe), Victoria (Hnyshe), and British Queen ; Beurre d'Amanlis may be used for the Jargonelle and Bergamots, as may also Bezi d'Antenay, the hardiest pear known. The cultivator has something to learn, for there are many pears of the finest quality, but of a delicate and infertile habit, that may be much improved by double grafting.

Our garden culture of cherries is, as yet, rude and imperfect; and espaliers of the Bigarreau and Guigne or Heart tribe, are planted and trained along the sides of the garden walks, giving abundance of shoots and leaves but very little fruit (which the birds appropriate), and in the course of time, give out. gum – owing to their having been unmercifully pruned – and die full of years and barren shoots, having given much trouble to the gardener. I have pointed out how cherries may be cultivated in gardens as pyramids, &c., and have alluded to fertility in the Bigarreau and Heart tribe being promoted by double grafting ; this mode of culture is also interesting as leading to success in soils that seem unfavourable to cherries under some circumstances.

Cherries grafted on the Mahaleb are described pp. 107 to 114 ; they affect calcareous soils, and, as far as I can learn, do not succeed so well in the sandstone formations, and where iron abounds in the soil; in such situations double grafted trees should be planted, formed in this way, – the common Morello cherry should be budded on the Mahaleb stock, and after two years it should be grafted with some kind of Bigarreau, Heart, or Guigne cherry; it will form a small or moderate sized tree, and bear abundantly. In cultivating cherry trees in soils inimical to their well-doing, abundance of chalk or lime rubbish should be mixed with the earth to the depth of two feet.

Double grafting of apples is of very inferior importance as compared with the same operation on pears or cherries, for our English Paradise stocks give the most perfect health and fertility in nearly all soils. Still there may be some peculiar positions where the soils are very light and poor, in which strong, robust sorts of the crab stock are required to make healthy fruitful trees. In such cases it is better to graft such sorts as the Hawthornden, Manx Codlin, and Small's Admirable, on thrifty crabstocks, and when two years old re-graft them with choice dessert kinds : all double grafting is best done when the first graft is two years old. I have now pointed out to a certain extent the advantages of double grafting, but much must be left to the intelligent amateur. It is to be regretted that English cultivators, more particularly nurserymen, have not turned their attention to the benefit choice fruit trees derive from having the proper kind of stock selected for them, or from being double grafted. Mr. George Lindley, father of the late Dr. Lindley, seems to have turned his attention to fruit tree stocks more than any other nurseryman of his day; still he knew only those grown by the nurserymen of his day, a very imperfect list. It is but a few years since that the common fruit-bearing quince, raised from layers – a most unfit stock – was sold for stocks for pears, and Mussell, White 'Pear Plum, Brompton, Brussels, and ' Commoners' (i. e. common plum stocks), are still the plum stocks propagated for sale ; all except the first and the last are of inferior quality and are surpassed by the White Magnum Bonum and the Black Damask Plum, which suit Peaches, Nectarines, Apricots, and all kinds of plums.

The double budding of some kinds of peaches and nectarines is almost necessary to their well-doing in some soils, yet this method of culture seems to have been neglected by European nurserymen. The truth must be confessed, that nurserymen, as a class, have but little taste for pomology ; they take to flowers and plants eagerly, because they give a quick return ; and thus Pomona and her gifts are always placed in theshade – as to experiments ' they do not pay.' There are some free growing kinds of apricots which, when budded on the plum, and the young apricot budded with a peach or nectarine, produce the most favorable effects on the peach tree, the union being perfect and the duration of it much lengthened. There are also one or two kinds of plums which, being budded on a wild kind of plum, form when budded with the peach or nectarine a most favorable stock, giving hardiness and fertility to the trees. We are still very backward in our knowledge of the effects of stocks on fruits : the subject requires much time and research, and no rushing to conclusions like some of our writers, who write on everything, and nothing well, only because they have not the necessary patience to master a few subjects thoroughly.

HOW TO PREPARE A PEACH TREE BORDER IN
LIGHT SOILS.

In our southern counties where light sandy soils abound, the difficulty of making peach and nectarine trees trained to walls nourish is well known; in spring they are liable to the curl and the attacks of aphides, in summer they are infested with the red spider, so that the trees are weakened, and rarely give good fruit: they seem, indeed, to detest light soils. The following method of preparing borders for them in such soils may be well known, but I have not seen it described by any gardening author. The idea has come to me from observing peach trees trained to walls, refuse to do well in the light sandy

soil forming a part of my nursery, except near paths, and to grow and do well for years in the stiff tenacious loam forming another part. My bearing trees in pots, for which I use tenacious loam and dung, rammed down with a wooden pestle, also bear and nourish almost beyond belief; and so I am induced to recommend, that in light soils the peach tree border should be made as follows: – To a wall of moderate height, say nine or ten feet, a border six feet wide, and to a wall twelve feet high, one eight feet wide should be marked out. If the soil be poor and exhausted by cropping, or if it be an old garden, a dressing of rotten dung1 and tenacious loam, or clay, equal parts, five inches in thickness, should be spread over the surface of the border ; it should then be stirred to two feet in depth, and the loam and dung well mixed with the soil. The trees may be planted during the winter, and in March, in dry weather, the border all over its surface should be thoroughly rammed down with a wooden rammer, so as to make it like a well-trodden path; some light, half-rotten manure, say from one to two inches in depth, may then be spread over it, and the operation is complete. This border must never be stirred, except with the hoe, to destroy weeds, and, of course, never cropped: every succeeding spring, in dry weather, the ramming and dressing must be repeated, as the soil is always much loosened by frost. If this method be followed, peaches and nectarines may be made to flourish in our dry southern counties, wherethey have hitherto brought nothing but disappointment.

1 If the border be new or rich with manure, a dressing of the loam or clay only, four inches in thickness, will be sufficient.

The two grand essentials for peach culture are stiff loam, or a very firm soil, and a sunny climate.

A CHEAP METHOD OF PROTECTING WALL TREES.

At Twyford Lodge, near East Grinstead, Sussex, the seat of B. Trotter, Esq., is a wall 75 feet long, covered with peaches and nectarines, which, for several years, had given no fruit; some years ago, the gardener, Mr. Mnrrell, asked my advice about protecting it with glass; and acting upon it with his own adaptation, has succeeded, every season since its erection, in securing fine crops of fruit of superior flavour. The following is a description of this simple structure: –

At the top of the wall, which is 12 feet high, is nailed a plate for the ends of the rafters to rest on ; 4 feet 6 inches from the wall is a row of posts, 6 inches by 4 (these should be of oak), 6 feet apart, and 3 feet 6 inches in height, from the ground; on these is nailed a plate to receive the lower ends of the rafters ; the latter are 8 feet long, 3 inches by !$, and 20 inches asunder; and the glass employed is 16 oz. sheet, 20 inches by 12. Every fourth square of glass at the top next the wall, is fixed into a slight frame of wood with a hinge at the top of each, and made to open all at once by a line running on a wheel; the front is of J-inch deal boards nailed to the posts, one of which, one foot wide, near the top, is on hinges, forming a drop shutter the whole length of the front. Now comes the management by whichred spider, the deadly foe of the peach tree, is discomfited ; and let me quote Mr. Murrell: –

' All these ventilators, back and front, I leave open day and night after May, except in very wet and rough weather. The first season I had the red spider (it was in the walls), but the fruit was of the highest flavour; the second season the fruit was very fine, and the spiders never came, I believe, owing entirely to my syringing the trees twice a day, morning and afternoon, and leaving all the ventilators open ; besides this the boards have shrunk, so that there are wide crevices, and the place is always airy. I thank you for your hints about giving plenty of air; the trees are admired by all who see them.'

The roof, it will be seen, is fixed, and the whole structure a fixture ; the trees can be pruned and nailed under shelter, and a crop of fruit always ensured ; how superior then is this to all the temporary protectors for walls so often recommended!

STANDARD ORCHARD TREES.

Although in this little work I profess to confine myself to the culture of garden fruit trees I feel that a few words as to my method of planting trees in an orchard under glass, may not be out of place, for very frequently a villa residence may have a piece of pasture land attached to it favourable to the growth of orchard trees, and quite necessary as a convenient place for the cow or the horse or horses. The common practice is to open large holes in the turf, six feet in diameter and from two to three feet deep ; and in the centre to plant a tree. In rich deeploamy soils trees often succeed when planted in this manner; and as often fail, the hole becoming in wet seasons a pond.

Orchard trees, as a general rule, should be planted twenty-four feet apart, row from row, and they are for the most part planted twenty-four feet apart in the rows, as to stand that distance apart over the whole orchard. I now propose that the rows should be twenty-four feet apart, but the trees twelve feet apart in the rows, so as to allow of one-third more trees to the acre. Instead of digging large holes, slips, six feet wide, should be marked out on the turf, so that the centre of each is twenty-four feet apart; each slip should then be trenched, or, as it is often called ' double-dug,' to a depth of two feet, turning the turf to the bottom of the trench and bringing the subsoil to the surface. A row of trees should be planted in the centre of each slip, twelve feet apart, and after the lapse of some fifteen or twenty years every alternate tree should be either removed and replanted or grubbed up. As such large standard trees would require much care in transplanting, and even then probably not succeed, the latter may prove the more economic mode. By thus planting more trees than required for a permanent orchard, a great advantage is reaped, for the temporary trees will, if the land is good, bear a large quantity of fruit, and amply repay their cost, which is trifling ; for whereas 95 trees are required to plant one acre, twenty-four feet apart, by the above method 142 may be planted. I have mentioned from fifteen to twenty years as the Trobable time when the temporary trees may beremoved ; as this depends entirely upon the quality of the soil and the progress they have made, a more certain rule to lay down is, that as soon as the outside roots of the trees touch each other the temporary trees should be removed. I need scarcely write the usual directions as to the trees being fenced round if horses and cows are turned into the orchard – that the trees should have stems at least six feet in height, and the lower branches should be taken off as soon as they become depressed enough for cattle to browse on them. One direction I feel, however, bound to give, – a circle from three to four feet in diameter round each tree should be kept clear of grass and weeds for at least five years from the time of planting, after that period grass may be allowed to cover all the surface as in old orchards.

In preparing the slips by trenching, if the subsoil be poor and stony, it should not be brought to the surface, but be merely turned over with the spade, and some manure mixed with it, keeping the turf, – well chopped – and the loose mould on the surface. If the soil be wet, drains four feet deep should be made twenty-four feet apart, one in the centre of the space between each row of trees ; they should be made with loose stones, which are far better than pipes for orchards. The bottom of the drain should be filled to the depth of eighteen inches with loose stones, and then filled in with the soil of the orchard. The soils best adapted for orchard trees are, first, loams with a subsoil of limestone; second, loams resting on a dry stony subsoil; third, loams resting on clay – these should be drained. Light L

sandy loams, with a subsoil of sand, chalk, and gravel, are not adapted for standard orchard trees unless the staple of loam is from three to four feet thick.

PROPER DISTANCES FOR PLANTING PYRAMIDAL AND OTHER FRUIT TREES.

Pyramidal pear trees and bushes on quince stocks to be cultivated as root-pruned trees for small gardens, four feet apart.

The same, in larger gardens, not root-pruned, six feet apart.

Pyramidal pear trees on the pear stock, root- pruned, six feet apart.

The same, roots not pruned, eight or ten feet – the latter if the soil be very rich.

Horizontal espalier pear trees on the quince stock, for rails or walls, ten feet apart.

Upright espaliers on the quince stock, for rails or walls, four to six feet apart.

Horizontal espaliers on the pear stock, for rails or walls, twenty feet apart.

Pyramidal plum trees, six feet apart.

Espalier plum trees, twenty feet apart.

Pyramidal and bush apple trees on the Paradise stock, root-pruned, for small gardens, three to four feet apart.

The same, roots not pruned, six feet apart.

Espalier apple trees on the Paradise stock, fifteen feet apart.

The same on the crab stock, twenty feet apart.

Peaches and nectarines, for walls, fifteen to twenty feet apart.

Apricots, for walls, twenty feet apart.

Apricots, plums, cherries, and apples, as single diagonal cordons, eighteen inches to two feet apart.

Cherries, as bushes and pyramids on the Mahaleb stock, root-pruned, for small gardens, four feet apart.

The same, roots not pruned, six feet apart.

Pyramidal cherries, on the common cherry stock, six feet apart.

Espalier cherry trees, for rails or walls, fifteen to twenty feet apart.

Vertical or diagonal single cordons of apples and pears, eighteen inches to two feet apart.

Proper distances for trees against dwarf walls, annually or biennially removed (see pp. 41 to 44) are for –

Pears on quince stocks, five feet apart.

Peaches, nectarines, apricots and plums, five feet apart.

Cherries and apples, five feet apart.

5

SECTION 5

MINIATURE FRUIT GARDEN CALENDAR.

January. – In mild weather, planting, root-pruning, lifting and replanting may be carried on. Some soils encourage the growth of moss on the branches of trees; lime may be sprinkled on them, or they may be painted with lime and soot formed into a thin paint with water.

February. – If the weather be mild, trees may still be planted without fear; the truth is, the modern system of growing fruit trees on dwarfing stocks, and removing them occasionally, makes them safe to plant very late or very early.

March. – If the clusters of blossom-buds are too numerous, they may now be thinned. Suppose there are five on a shoot six inches long, two of them may be removed ; and if five are in a *duster,* three should be cut out; if seven, four, and so on. This is quite necessary with apple trees on the Paradise stock and pears on the quince stock, as they are often too much crowded with blossom-buds in clusters of from five to ten and upwards; a small sharp knife should be employed in this operation. Towards the middle of the month protecting (see pp. 26, 44) to retard the blossom-buds is good practice. Planting of prepared or oft-removed trees may still be safely practised.

April. – Protecting (see p. 27) should still be attended to. It would be an innocent and perhaps useful experiment to sprinkle on the cluster of blossom-buds, just on the point of opening, when they are moist with dew or gentle rain, a coat – say one-eighth of an inch thick – of cocoa-nut fibre ; this will protect them from still hoar frosts. When washed off by heavy rains it may be renewed.

Planting of pears on quince stocks ; the buds on the point of expansion (see p. 60), may be tried as an experiment: here they often bear the finest fruit.

May. – Towards the end, if the season be early and the young shoots have made from five to six leaves, they may be pinched (see pp. 9, 12, 49).

June. – Summer pinching must be strictly attended to; the young fruit in clusters should be thinned, removing from pears and apples about half their number.

July. – Summer pinching still to be attended to; the very early kinds of pears should be gathered before they are quite ripe.

August. – The leading shoots of the lateral branches of pyramids (see p. 10) should now be shortened, and towards the end, early ripening pears – viz., sorts that ripen in September and early in October – may be gathered, unless the season be late.

September. – Shortening the shoots if omitted in August, may be done; gathering of early pears before fully ripe to be attended to. Towards the end gather apples and pears that ripen before Christmas.

October. – Towards the middle of the month 140 THE MINIATURE FRUIT GARDEN CALENDAR.

planting may be commenced; and if the rain has penetrated sufficiently, root-pruning may be- done ; also lifting and replanting (see p. 125). About the middle gather late pears.

November. – Planting', root-pruning, lifting, and replanting may still be safely carried on.

December. – All the operations of last month may still be practised if they have been forgotten or neglected.

Always bear in mind that a vigorous growing tree, that does not bear fruit, requires being lifted and replanted – even annually – till it becomes fruitful, and that a tree that bears well and makes annual shoots under twelve inches in length, requires neither root-pruning nor removal, but merely summer pinching of its shoots to about half their length.

6

SECTION 6

APPENDIX.

THE GROUND VINERY.

The ' Curate's Vinery,' described in the tenth edition, was contrived by Dr. S. Newington, of Tice- hurst, and consisted of a ridge of glass placed over a furrow lined with slates, so that the bunches of grapes were suspended in the furrow, and in warm seasons ripened well. One objection to the furrow was its liability to be filled with water in wet weather, in low situations and heavy soils. I therefore sought to remedy this, and one day, about the end of June, 1860, I found myself looking into my original ' Curate's Vinery,' and admiring the vines then in blossom, although those within a few yards of it, growing in the open air, were scarcely in full leaf. I pictured to myself the bunches of grapes suspended from the vines in the warm, moist atmosphere of the trench lined with slates. My thoughts then reverted to my boyish, grape-loving days, when in an old vineyard, planted by my grandfather, I always looked for some ripe grapes about the end of September ; and I vividly remembered that I always

found the best and ripest bunches with the largest berries lying on the ground, and if the season were dry and warm, they were free from dirt and delicious, and so I gradually travelled in thought from bunches of grapes lying on the ground to *idem* lying on slates.

The idea was new, and I commenced at once to put it into practice by building a ' Curate's Vinery ' on a new plan.

I therefore placed two rows of bricks endwise (leaving four inches between each brick for ventilation) on a nice level piece of sandy ground, and then paved between them with large slates (' duchesses') placed crosswise. I am, however, inclined to think that tiles may be preferable to slates ; absorption of heat is greater and radiation slower. On the bricks I placed two of the ridges of glass, as given in the foregoing figure, each seven feet long, and thus formed my vinery, fourteen feet in length. The vine lies in the centre of the vinery, and is pegged down through the spaces between the slates. One vine will in the course of two years fill a vinery of this length ; but to reap the fruits of my project quickly, I planted two vines, one in the centre, the other at the north-east end, for these structures should stand north-east and south-west. One of these vines, which had been growing in a pot in the open air, was just beginning to show its fruit-buds – it was quite the last of June – its fruit ripened early in October, and were fully coloured and good in spite of the cloudy cold autumn. My black Hamburgh grapes in my ground vineries were fullyripe in 1862 by the first week in October. I therefore feel well assured that grapes lying on a floor of slates such as I have described, will ripen from two to three weeks earlier than in vineries of this description with a furrow, and as early as grapes in a common cold vinery. Black Hamburghs, and other kinds of grapes not requiring fire heat may thus be grown in any small garden at a trifling expense. I am, indeed, disposed to hope that the Frontignans, and nearly all but the Muscats, may be ripened by this method, so intense is the heat of the slated floor on a sunny day in July.

Some persons may think that the heat would be scorching, and that the leaves and grapes would alike become frizzled ; but few gardeners know the extreme heat a bunch of grapes can bear. I remember a lady friend who had resided some time at Smyrna, telling me that one afternoon at the end of summer, when the grapes were ripening, she was sitting in her drawing-room and admiring some large bunches of grapes hanging on a vine which was growing against a wall in the full sunshine. Knowing the danger of going into the open air without a parasol, she rushed out, cut a bunch of grapes, and returned to her seat in the shady room. The bunch of grapes was so hot that she was obliged to shift it from hand to hand. I observed in the hot weather we had in July, 1859, one or two branches of Muscat grapes nearly touching the chimney of a stove in which a fire was kept up every morning, gradually turning into raisins. I felt some of them when the sun was shining onthem, they were not burning hot, but next to it. I allowed them to dry into raisins, and very fine they were, but not better than the finest imported from Spain.

With respect to the superior ripening power of slates or tiles placed on the surface of the earth, I was much interested in once hearing a travelled friend say that when he was at Paros, he observed many vines trained up the marble rocks peculiar to the island; and in all cases the grapes lying on the surface, which was almost a continuous mass of rock, were ripe, while those a few feet from it, on the same vine, some of the branches of which were trained up the wall-like rocks, were quite green. In telling me this, he said he was never more impressed with the ripening power of the earth's surface.

I have, in giving the figure and description of the ground vinery, adapted for one vine, the width of it being 2 feet 6 inches only. If this width be increased to 3 feet 6 inches, two vines can be trained under the same roof 14 inches apart, and thus at a trifling additional cost double produce can be obtained.

- Cultivators will think of red spider making his home in such (for him) a happy, hot place; but he may be made so uncomfortable by keeping flowers of sulphur strewed over the slates till near the ripening season, that no inconvenience need be apprehended. It will be perceived that the ventilation is all lateral, and on the same principle as that of my orchard- houses ; nothing can be more perfect. In the figure it will be seen I have left a small aperture under the apex of the roof for the escape of rarefied air. In very hot weather this may be useful, but in my slate-floored ground vineries I have not done this, and yet the ventilation is perfect. I have not yet ascertained in what manner the heated air escapes. The ventilating apertures are all on the surface of the soil, and at the same level; but I suppose it stoops to get out, having no other mode of egress.

DIMENSIONS OF GROUND VINERIES.

No. l, for a single vine in centre,
Width at base 30 inches.
Slope of roof 20 inches.
Depth in centre . . . . . .16 inches.
No. 2, for two vines 14 inches apart.
Width at base ...... 42 inches.
Slope of roof 28 inches.
Depth in centre 20 inches.

These dimensions need not be arbitrary, for ground vineries of larger dimensions may be made with every chance of success, and Hamburgh grapes grown in Bedfordshire instead of cucumbers ; for no part of England can be more favourable to grape culture than the fertile, sandy districts of a portion of that county. We have heard of forty acres of cucumbers being grown for pickling, and one day we may hear of forty acres of grapes in ground vineries in some favourable locality. To form a vinery (p. 142, Fig. 22), described above as No. 1, two seven-feet lengths are required; these I find from experience are better made of wood than iron, which is heavy and expensive; they are now made three feet wide at base, and sold by Mr. J. Rivett, Stratford, Essex, unglazed and unpainted. Their size may also be increased to 3 feet 6 inches, aS described under No. 2, but they must then be placed on a wall two bricks in height, leaving apertures, four or five inches wide and six inches deep, for ventilation; this increase of ventilation is absolutely necessary with No. 2. The glass used should be 21 oz., as 16 oz. is too slight. As the vines in ground vineries often put forth their young shoots early in May, and are apt to be injured by a severe May frost, it is good practice to keep some refuse hay strewed over the glass when there is any chance of frost in that month, or to cover the ridges with mats. In gardens where these glass ridge roofs are not wanted for vines or fruit tree culture, they will be found most useful. They may be placed on any warm border on bricks; and early peas, French beans, and many other early vegetables requiring protection from spring frosts, be grown under them with advantage. For the cultivation of early strawberries they are invaluable, as they not only hasten the ripening period, but protect the fruit from heavy summer showers, often so injurious to the crop, and also from birds. Strawberry plants, to be cultivated in ground vineries, should be planted early in autumn in narrow beds of two or three rows, the plants close together in the rows, so as to take full advantage of the glass-covered space. The rows should be 9 inches apart, and the plants in the row the same distance from each other; the beds shouldbe made every season on a *fresh* piece of rich soil; and as much fruit as can possibly be grown in such a limited space must be the aim of the cultivator. If the ridges are devoted to strawberries only, much care is required in their culture, the runners should be carefully removed and the glass ridges taken off after the fruit is gathered, and not replaced till November; the plants will require water and surface manure during the summer. In all cases the ridges should be placed on bricks, with spaces between them. Ventilation is then secured; and even cauliflower plants in winter will do well without the constant attention to 'giving air,' so necessary in the old garden frame culture. Lettuces, for early salads, succeed admirably in these structures ; they should be planted in October. In gardens that are confined and very warm, I repeat it may be necessary to have a small opening left at the top, at *a* in the figure, just under the ridge, to let out the heated air, and two rows of bricks instead of one ; but my vineries stand in a very exposed place, and do not require it.

MANAGEMENT OF THE VINES.

It is now (1870) twelve years since ground vineries were invented, so that time enough has elapsed to know their utility or the converse. My vines are now from nine to ten years old, and as it may interest distant readers, I will endeavour to describe them : –

My oldest and finest vine is the Trentham Black ; this occupies seven eight-feet lengths, and is 56feet long; this put forth a vast number of bunches this season, of which about 100 are left to ripen their fruit. My second in age is a Black Hamburgh – the variety called Belle Bruxelloise in Belgium, it is a great bearer and a little earlier than the old sort; this runs through four seven-feet lengths, is consequently 28 feet long, and is loaded with fruit. My third vine is the Buckland Sweetwater, which occupies four seven-feet lengths, and is also full of fruit. These three vines have never failed in giving and ripening nice crops of fruit, ever since they were planted: there are many other kinds of grapes cultivated here with success under these simple structures, but the above are the oldest and best established. I may add that it is as yet difficult to place a limit to the growth of a vine under a ground vinery if the soil be favourable, viz., calcareous sandy loam, or even calcareous clay if well drained. I fully believe that a planter not too far past his fourth decade, may live to see his vine 200 feet long, and covered with fruit from end to end – the artificial climate created by the glass ridge seems so highly favourable to the development of the plant. There is a most essential rule to be observed, – *the vino must be covered with its glass ridge all the year round,* with the exception of a week or two in autumn.

PLANTING AND PRUNING VINES FROM POTS.

The most preferable seasons for planting vines from pots are in October and November or in March, the latter to be preferred, and if vines can be placedin a cold vinery or under a garden frame till their young shoots are two inches long, they had better be planted in April, as they seem to start with greater freedom when their young shoots have commenced to grow. The mode of planting as practised here is simply to mark out a piece of ground 3 feet square at the end of Fig. 22, and to dig it 2 feet deep, mixing with the soil, in digging, a coat of manure from four to five inches thick, placed on the surface before digging; the vine should not be planted under the glass, but outside, at one end, it should at once be pegged down with two or three hooked pegs thrust into the earth through the interstices between the slates in the centre of the floor. If vines from the open ground are selected, they should be planted early in March, and cut down to two eyes ; if strong vines from pots are planted, they should have their roots carefully divided and spread out; to do this the ball of earth should be squeezed between the hands so as to loosen it thoroughly, and after planting, water should be given, the earth filled in, and after about ten days the soil round the vine should be trodden firmly ; the vine from a pot, if strong and from 7 to 9 feet in length, should be shortened down to 3 feet, or say, to 11 or 12 buds, not counting the buds within 9 inches of the ground ; every bud will show a bunch of fruit; all but three or four bunches should be removed, and every side shoot except one should be shortened as soon as it has made, say, five leaves ; the one to be excepted is the leading shoot, which if the vine is growing tolerably well, may be suffered, even the first season, to growfrom, four to five feet before it is stopped: this leader may require being stopped a second time the first season if it is in a vigorous state. In the autumn (mind this is the first season) the young leading shoot may be cut down to about twelve eyes, or within three feet of the old wood, *i. e.*, the shoot left on the vine when planted, the latter will be furnished with spurs and each of these must be shortened in the autumn to two eyes, the time for pruning is towards the end of October; after the fruit is gathered, and at this time only, the ridges may be removed from the vine, and remain off for a fortnight: the pruning in succeeding years is very simple, you have merely to shorten the leader to three or four feet, or less, and the spurs to two eyes annually in October.

During the winter, if the vineries are standing in an exposed situation, they should be secured from the wind by driving a few stakes down on each side. In spring, if the vines put forth their young shoots in April, they are apt to be killed by a spring frost, as is too often the case with the vines of France ; this can, however, be easily averted in ground vineries either by keeping constantly their covering of hay or straw on the glass when the weather is cold, and frost likely, or to cover the ridges with the small mats which are so convenient and so cheap, whenever the. thermometer declines to 40 at 7 P. m.

'there are still more ills to guard against in ground
vinery culture, for mice and birds, as rats often do in
common vineries, attempt to have too large a share
of the fruit; they enter by the interstices between
Mthe bricks and devour and spoil many bunches ; thrushes are particularly vigilant in looking after grapes, and may be trapped, but both they and the mice may be kept out by galvanised iron netting six inches wide, placed along the whole length of the vineries.

I have but little to add to my description of the management of ground vineries: their uses are endless, for not only are the finest of pears grown in them, but peaches, apricots, plums, and strawberries may be cultivated with great success, and then as winter quarters for bedding plants they are excellent; for this purpose the bricks should be removed in severe weather, and the glass ridges thickly covered with straw, they are then perfectly frost-proof; in mild weather in winter the ventilating bricks may be replaced, and the straw removed till frost again occurs.

With respect to the most preferable dimensions for these structures – the size No. 1, thirty inches wide at base, will suffice for one vine in the centre *j&r* ten years or so ; but as I perceive my old vines to be a little straitened for room, I advise a width of three feet at the base, and No. 2, for two vines or two cordons, of three feet ten inches, instead of three ieet six inches.

in these more roomy structures the vines may be trained to stout galvanised iron wires, supported with iron rods flattened at top and perforated, so that the wire passes easily through ; these wires should be about one foot from the surface of the slates, and the suspended bunches, partially rest ing on them, will ripen admirably. I ought to add, that a friend with much gardening experience finds his strawberries ripen ten days earlier than those in the open air, and his melons, planted on new, fresh, fermenting manure, in a trench, are free from red spider, and produce fine fruit. It is the constant ventilation, night and day, and the heavy dew, the result of arrested radiation, that seems to baffle this tiresome plague, for although my vines are never watered or syringed, they are always vigorous and free from red spider. The most eligible varieties of grapes for ground vineries are, the Black Hamburgh, Buckland Sweetwater, Royal Muscadine, Early Smyrna Frontignan, Trentham Black, Early Saumur Frontignan, Meurthe Frontignan, and Esperione.

Any suburban garden ten yards square, if in a sunny situation, may have one or two of these vineries, and the owner or occupier may grow his own black Hamburgh grapes, known by most of the Londoners as ' hothouse grapes.' I ought to mention that the improved ground vinery, with one side on a hinge, so that the side opens and gives access to the interior, is the best of all, and is made by Mr. J. Rivett, Stratford, Essex.

CORDON TRAINING.

Py T. Francis Rivers. Extracted from the 'Journal of Horticulture,' Nos. 356-7. The introduction of the system of training fruit trees, called by the French, cordon training, leads
"

me to suppose that a few outlines of description may not be unacceptable. This system of training is remarkable for simplicity, and I propose to give the necessary directions in as few words as possible.

The preparation of the ground is so well understood, that is not necessary to say much on this point. To form the oblique-cordon orchard, a trench should be dug, about two feet wide, the first spit of soil being thrown out as if for a celery trench ; the under spit should then be broken up and left with the top soil, a good proportion of well decomposed manure must be mixed, and the ground is ready for planting. The trench should, if possible, be made about a fortnight before planting, in order that the soil may be thoroughly pulverised. If there is any deficiency of lime in the soil, it is as well to add lime rubbish or chalk. For horizontal double cordons a trench is not necessary; holes should be dug about two feet in diameter, and the soil mixed with good compost. The double-cordon trees should be from twelve to fifteen feet apart; the horizontal single cordons six to eight feet. At this moment there are at Sawbridgeworth two horizontal double-cordon peach trees under a ground vinery, which measures twenty-one feet from end to end, and promise, from their remarkable vigour, to be models of cordon culture next year, every spur being full of strong fruit buds.

Fig. 1 respresents a double horizontal cordon. This may be made by cutting down a dwarf maiden tree to within four or six buds of the base, the two topmost buds of which must be selected to form thecordons. The highest on the stem are the most eligible; but the operator can, of course, select the two shoots which are the most convenient for his training wire, and they should be as nearly as possible opposite. When sufficiently advanced in growth to be flexible, they should be carefully bent down and fastened to short sticks, unless the training wires are used. As the whole energies of the tree are directed into these shoots, they will make rapid growth, and as they advance, fresh sticks and fresh tying will be necessary. As any lateral or upright shoots are put forth they must be stopped at three or four leaves from their bases. The first year few of these will be made, but the tree will most probably, if there is a favourable growth, be studded with fruit buds. In November, or, indeed, any month from November to March, the tips of the main shoots should be shortened three or four buds from the ends, and unless a few lateral shoots have been left, which should be removed, the pruning for the first year will be accomplished.

The second year each cordon, or branch, will produce many lateral shoots, and as these are successively produced they should be pinched. The first pinching must be done when the shoot has formed five or six leaves, and as a general rule, three leaves from the leaflets should be the stopping point. This primary shoot will form the bloom-buds, and the shoot made from the terminal bud must be stopped in the same manner as the first. Discontinue pinching after August. By this time the cordon will be thickly studded with wood and fruit spurs; to thin

out and regulate these will form a pleasant winter morning's work ; the final pruning must therefore be deferred until November.

The tree after the second year will assume the appearance of a cordon – i. e., a thick rope of closely studded shoots, and the pruning must be left to the judgment of the operator. Many shoots must be removed ; and as the size and strength of the tree must regulate the number of fruit-bearing spurs, a sufficient number of these being left, the operator should prune all others to wood-buds, in order to produce, year by year, an alternate succession of fruit-bearing wood.

Fig. 2 is a half-standard double horizontal cordon. This is very useful for low walls in gardens ; where the border is occupied by flowers or other plants, the part of the wall exposed to the sun may thus be used. A standard cordon with a stem six feet high may also be used for the top of the wall, the main surface being occupied by other trees. A cordon fringe, or cornice, will be found exceedingly ornamental, and may be carried the entire length of a wall, the standards being planted at intervals of twenty feet or more.

Many other forms of cordon training will, doubtless, be discovered as the system becomes better known.

Single horizontal cordons (Fig. 3), require -the same pruning as the double, but the dwarf maiden tree does not absolutely require the cutting-back necessary for double cordons. The tree may be planted in a slanting position against the training wire, and the shoot tied down. The first year after planting most of the bnds will break and produce

shoots ; these must be treated in the same manner as the double horizontal cordons. If a single cordon is required for a special height, the shoot shonld be shortened to the height required, and a single horizontal shoot selected to form the cordon.

Single oblique or diagonal cordons may be planted to training wires by the sides of walks, or in rows in the garden devoted to their cultivation. The space given up to them will yield an ample and quick return in fruit. They may be planted 1 feet apart, and if the cultivator does not object to wait a year, dwarf maiden trees are the best to plant, as they may be bought cheaply. The trees should be planted upright, and the shoots, which are generally very flexible, should be bent to an angle of about 45. It is not necessary for the angle to be quite acute ; but, as a general rule, this angle may be adopted. If the shoots are not flexible enough to bend, plant the tree in a slanting position.

The principle of pruning given for double horizontal cordons must be followed in the cultivation of single oblique cordons. They will the first year after planting be found covered with bloom-spurs. Single oblique cordons in rich and fertile soils will, probably, require root-pruning as well as spur- pruning and, if necessary, this should be done every second year. The tree should not be taken up, but the spade pushed down at a sufficient distance from the stem to avoid injury to the main roots, and the tree gently heaved. If a tap root has been made it should be cut. The proper time to perform this operation is near the end of October, and any time afterwards to the middle or end of February; but it is better done in October and November, as many *fresh* roots will be formed after the operation, even

during what are called the dead months of the year.

Single oblique cordons may be carried to the

height of ten or twelve feet; in fact, there is no limit,

A Photograph of Doyenn6 de Cornice.

Mg. 4.

except the will of the planter. A fresh string of wire may be added annually as the cordons increase in length. They may also be limited to the height of four or five feet.

Fig. 4 is from a photograph of an upright trained tree, with five vertical cordons springing from a common base. Trees may be purchased already trained in this form, but the double horizontal cordon may at pleasure be changed into this form by selecting strong shoots at regular intervals, fastening them to stout stakes, and summer pinching them as practised for oblique cordons.

Fig. 5 is a fan cordon, and the advantage of the simple method of summer pinching will at once be seen. Instead of a wall being perforated all over with nails, few only are required to fasten the shoots selected for cordons. This form may consist of five, seven, or more cordon branches. The symmetry of the tree should be the point most strictly attended to, a symmetrical tree being more pleasing to the eye than one irregularly shaped. The same method of pruning is required as for oblique cordons.

Fan cordons can be managed by an unscientific gardener, but to produce one well-shaped on the usual plan requires a skilful and practised hand. It is possible that in the northern and westerly districts peach and nectarine trees will produce too many unripened spurs, but probably by attention and strict thinning this difficulty will be surmounted. It is not yet sufficiently known that apricot fan-shaped trained trees may be made by the most simple management of cordon training most prolific and easily managed wall trees. The method is this : – as soon as the tree has formed a perfect-shaped tree, no more shortening of shoots or ' laying in' of young should be practised, but every branch should be

Fig. 5. made into a cordon by summer pinching, *i. e.*, nipping off early in June every side shoot to 4 or 5 leaves, leaving the end of the cordon shoot untouched till, say, February, when, if it be more than 30 inches in length, it may be shortened to 20 or 24 inches. Peaches, nectarines, and all other kinds of wall

Fig. 6.

fruits may be grown after this cordon system, and if the walls be not very extensive, much room may be saved by adopting the five-branched upright cordon (Fig. 4, p. 161). A clergyman in Yorkshire, a clever cultivator, asserts the latter as the best of allpruning for all kinds of fruit trees for furnishing walls, for it must be recollected peaches and nectarines may be cultivated after this method, and a great variety of fruit obtained from a small garden. With peaches and nectarines in rich soils it may be necessary to leave one shoot on each branch as an exhauster – an unpruned shoot – or to lift the tree once in three or four years. I can only assert that this system of training fruit trees against walls is so agreeable and so nice, that one becomes quite attached to it. I should add that the exhauster should be cut down in winter to three or four buds.

Fig. 6 is a double oblique cordon, formed by cutting down the dwarf tree to two buds, and proceeding as for oblique cordons.

Fig. 7. represents a compound horizontal cordon. This should have a central shoot and branches trained from it as nearly opposite as possible. This system has long been used for pears and apples, but not so generally for stone fruits. It is well adapted for peaches, nectarines, apricots, cherries, and plums. All of these may be trained as compound horizontal cordons in the colder climate of Yorkshire.

A very skilful cultivator of fruit, in Yorkshire, has trained cordon peaches and nectarines with complete success, and to counteract the tendency of these fruit trees to produce much unripened wood, when under cordon training, he leaves on every horizontal branch an upright shoot which he calls an exhauster. This shoot forms an outlet for the superfluous energy of the tree ; and the fruit spurs, being deprived of the superabundance of the vital fluid, do not break into growth. This theory will be found to be very sonnd practice, and should be used wherever there is a tendency on the part of the tree to

produce many uuripened spurs. This mode of training for the pear and apple is already well known; and when applied to peach and nectarine trees, the only deviation from established practice will be to treat every horizontal branch as a cordon, and to practise summer pinching instead of allowing gross upright shoots to be made.

Fig. S.

Fig. 8 is a single vertical cordon in a pot, and if an orchard house or glass shed is available, these will be found very useful and interesting trees. Pear, apple

cherry, and plum trees may be potted into 10 or 12 inch pots, and moved into a glass shed, or, indeed, any shed open to the sun, while in bloom, and kept under cover until all danger from spring frost is past. They should then be removed to a border prepared for them – the warmer and more sheltered the better. The pots must be plunged to within 2 or 3 inches of the rim, stable litter partly decomposed and spread over the pots and the soil; as the trees will require watering, they should be placed near water. One- year-old dwarf trees may be bought at a cheap rate and potted. The fruit will be produced in the second year after potting. The soil for the , trees should consist of good, strong, calcareous loam mixed with a third of its bulk of decomposed manure. An old cucumber or melon bed may be used ; or if not convenient, stable manure thrown up and fermented for some time will answer very well. The soil must in all cases be made very firm and solid in the pot. The border or bed for their summer quarters should be 6 feet wide ; this will take four rows of trees. This distance is perhaps the most convenient for pruning and watering, but it may be increased or diminished at the will of the cultivator. ,

Under this system trees which appear to be walking sticks in the winter will become wonderfully fertile ; and if protection in spring can be afforded, the crop is almost certain. As it is possible and probable that daring the summer some of the roots will have passed through the bottom of the pots inte the soil beneath, it will be necessary, after the fruit is gathered and the trees are at rest, to detach them from theiranchorage by taking up the pots and cutting off all the roots that protrude through the drainage hole of the pot. As this operation will break up the summer quarters of the trees, there will be no necessity to replace them at the distance requisite for their summer cultivation. They may be much more closely packed for their winter quarters, plunging them as mentioned before, and during winter covering the pots thickly with straw or stable litter. In this position they may be left without any further care or attention until the returning spring urges them again into fresh activity and fruitfulness.

The Compound Trellis. See diagram, p. 170.

The end posts A A, 3 ft. by 5 in., of oak, 5 ft. 6 in. out of ground, and 3 ft. in ground, with blocks 2 ft. long, D, and brace c to take the strain, with four rows of No. 13 galvanised wire strained by Raidis- seurs, the first 14 in. from ground, and 1 ft. apart. The top No. 7 strand wire is 4 in. from top, strained by screw.

At each end post a brace, 4 ft. by 4 ft., is fixed, 5 ft. up and 5 ft. from post, with intermediate stays of iron, E, T3ff in. by 1. in., in these there are six rows of No. 13 wire, 1 ft. apart, and strained the same way as the upright wires.

Middle tier, H, of posts, 7 ft. 6 in. out of ground, and 3 ft. in ground with stiff brace, are fixed about 35 ft. apart, with five rows of No. 13 wire, first row 1 ft. 4 in. from ground, three others 1 ft. apart, and one between them and top strand, which is No. 7 and fixed 4 ft. from top of post.

Length 200 ft., width 16 ft.

The plan of the compound trellis will, I hope, be understood by my readers. The two'diagonal trellises D D and the outer trellises A A should be planted with upright cordon trees (see fig. 7). The centre trellis H may be planted with fan-trained trees, either plums or cherries, or, in favourable climates, with peaches, nectarines, and apricots. I propose to plant fan-trained trees on the centre trellis because they may be planted at some distance apart, 20 feet; the diagonal and the outer trellises I propose to plant with plums, pears on the quince stock, and apples on the Paradise stock.

The main object of a trellis of this kind is; of course, to afford protection during the spring, and this, I think, may be given in various ways. The stout wire on the top of the posts is intended to support either mats or canvas; as the protection is for a time only, mats will probably be the cheapest. If the protection is intended to last for some years, painted canvas will answer. As protectors for the outer rows of trees, I think straw mats, or hurdles covered with straw, will be the most economical.

For market purposes I should recommend planting Doyenne du Comice, Louise Bonne of Jersey, Duchesse d'Angouleme or Easter Beurre, and possibly other large sorts of pears ; with good cultivation profitable results may be realised, the first outlay not being very considerable.

INDEX.

The following Works by Mr. RIVERS are sold by Mrssrs.
LONGMAN & Co., or sent free per post, at the prices quoted, on
application to the Author, Sawbridgeworth, Herts.

THE HOSE AMATEUR'S GUIDE;

Giving the History and Description of the finer kinds of Roses ; with
Directions for their Culture in the open Air and in Pots. Ninth
Edition, just published. 4s.

THE ORCHARD-HOUSE ;

On the Culture of Fruit Trees in Pots under Glass. 15th Edition.

THE MINIATURE FEUIT GAEDEN ; .

On the Culture of Fruit Trees as Pyramids and Bushes., 17th

Edition. 3s.

A DESCRIPTIVE CATALOGUE OF FRUITS ;

Carefully compiled and arranged, so as to be a complete Guide to the Purchaser of Fruit Trees. Free.

A DESCRIPTIVE CATALOGUE OF SELECTED ROSES.

Free.

THE CONGRESS PAPERS;

Containing Orange Culture, &c. *It.*

Spottiswoode fr Co., *Printers, Jfeif-Hreet Square, London.*

Lightning Source UK Ltd.
Milton Keynes UK
24 June 2010